OUR SIX DIMENSIONS

THEIR DISCOVERY, TESTING, AND MEANING

DR. STEPHANIE LAWRENCE

Our Six Dimensions
Morrison

OUR SIX DIMENSIONS®

THEIR DISCOVERY, TESTING, AND MEANING

Library of Congress Control Number: 2019910044

ISBN: 978-1-733-32570-7 (sc)
ISBN: 978-1-733-32571-4 (hc)
ISBN: 978-1-733-32572-1 (e)

Printed in the United States of America

I happened upon an interesting idea and

now I share the idea.

CONTENTS

Preface   ii

Chapter 1   Five Plus One Equals Everything   **1**
My Soapbox   2
Let's Start the Debriefing   9
Starting to Collect Business Cases   12
My Research Perspective is Updated   13
Figuring Out the Puzzle   16
Scales and Increasing/ Decreasing Values   20
Six Dimensions in Real Life   22
Everything is Constant, But Some Constants
    Have Names   24
Right, But How Do You TEST Dimensions?   27

Chapter 2   I Used Research Papers as My Data Set   **31**
Dimension's Test Takes Shape & More
    Business Cases Included   32
Stratified Proportionate Sampling Using
    Peer Reviewed Papers   39
Testing Results for Dimensions   42

Chapter 3   Let's Talk About Theory of Businesses as
Urban Cultural Icons   **47**
Theories Contributing Towards Businesses as
    Urban Cultural Icons   48
Sociology v Geography's Icons plus
    Totemism   51
Urban Theories Begin Mentioning Dimensions   53
Businesses Viewed From Urban Perspective   57

Chapter 4 Let's Start Measuring Businesses   **61**

Businesses Added to the Data Set   63

Overall Scale Formatting for This Research   64

Starting Guttman Scale by Measuring Entity   67

Measuring Place for Urban Cultural Icons   72

Scaling of Time for Urban Cultural Icons   75

Business Transaction Categories   80

Theming and Product Line Extension
  Categories   83

Vice Categories of Businesses   88

Chapter 5 Measuring, Explaining, and Diagramming
            the Six Dimensions   **93**

Hypercube Roles   95

Measurement Examples to Clarify Each
  Dimension   97

Chapter 6 Tehran's Grand Bazaar is the Most Local
            Cultural Icon   **109**

Adding Up Business's Numbers   113

Local, Focal, Zonal, and Global Cultural
  Icons   120

Weaknesses in this Research   129

Changes Since This Research   130

Confirmation We are Measuring the Right
  Stuff   133

Chapter 7 What Aristotle and Buddies Bring to the
            Table   **141**

Strengths and Weaknesses of Other
  Perspectives   142

Multiple Realities and False Dichotomies   145

Linear Sequences   149

Hot Spot #1: Negative Time   153

Hot Spot #2: Energy Plus Consciousness   154
Hot Spot #3: Communication   156
Hot Spot #4: Language   158
Hot Spot #5: The Six Dimensions   160
Theory of Everything and Unified Fields   165

References   **171**

About the Author   **189**

Preface

During my final reading I realized people might not have the same background knowledge as my background, so on www.oursixdimensions.com I have listed links to discussion topics. Overall this is not a big and long discussion where concepts and layers are pontificated forever and you ask when does this agony end. I believe using regular words as much as possible.

Our Six Dimensions is a direct examination of a dense topic. I volleyed back and forth that this book is not clear enough or specific enough, but that is a false statement. These dimensions convey insights and uses on multiple levels and by way of being provided with the grand perspective you should have an Aha moment. Afterwards you can look up and review topics that were a little fuzzy to you.

The Six Dimensions are the star of this book and because of this insight Chapter One is a bit dry because I am setting the stage and introducing characters. I am also tight when it comes to using adjectives which means I wanted to bland out secondary topics to allow the dimensions to shine. This means while there are back stories with conversational threads included in the book, I did not want to sway the conversational focus.

I never expected to identify these dimensions, but I observed them and am tasked with explaining these dimensions to others. Along this same line, this new insight will cause a shift in perspective. You are tasked with putting these dimensions and new insights into your life skills toolbox and using them to accomplish tasks and become empowered. Be intuitive. When you read specific examples, then generalize concepts to your life. Start looking for similarities and differences, disassemble the process mechanics, and look for outliers. For academics your research paradigm will expand and be stronger.

A buddy asked 'How did you see this? How did you think this far outside of the box?' The only thing I can think of is I connect ideas together. Weiner dog Roxie was chewing on a treat this morning and I video recorded her in action with my running monologue saying 'She's like a typewriter going across a row and then starting the next row of nibbling.' Thank you for being curious, interested in learning something new, and willing to change.

1

Five Plus One Equals Everything

I discovered *Our Six Dimensions* while researching a method to format a scale for my dissertation in order to measure the cumulative iconic value of brick-and-mortar businesses in society. This explains why businesses as urban cultural icons are such an integral part within the discovery of our six dimensions. I have also presented my research findings at conferences and when I look back at the history of *Our Six Dimensions* I am still amazed this entire grand idea of everything came about from me wanting to see if I could measure businesses as urban cultural icons and being curious as to the sum scores which would occur with various businesses.

Because businesses as urban cultural icons begat the six dimensions this book has two topics, the six dimensions plus measuring businesses as urban cultural icons. Both are intertwined plus occasional sprinkling of theories and stories to support topics being examined. In addition to including these six dimensions into your skills toolbox also listen to included stories and insights plus the ones you will hear through life. They are your clues towards problem solving.

When I was chumming for the creation of my dissertation topic I saw the rawest form of our six dimensions. It is now 15 years after the publication of *The Globalization of Nothing* (Ritzer 2004) and during this time no one else has presented or published anything about the six dimensions. This makes me wonder how many bazillion times in the past 5,000 years have humans stood in this discovery only to walk away from this diamond in the rough. We see, interact, and identify everyday with these dimensions and now is your Eureka moment. When you buy groceries, watch a movie, read a book, ride a bicycle, or think about things our six dimensions are right there with you, every time.

My Soapbox

Before Hurricane Katrina came through New Orleans I walked into University of New Orleans (UNO) bookstore because I was hunting for inspiration for my dissertation research topic. That spring I was going to create my dissertation topic and I was willing to research pretty-much any topic. I have always liked George Ritzer's writings because he creates plus expands interesting theories about society. My research path started innocently enough when I picked up George Ritzer's *"The Globalization of Nothing"* (2004) at UNO's campus bookstore.

Prior to his 2004 book in *McDonaldization of Society*, Ritzer (1993) wrote about how jobs are becoming compartmentalized by dividing tasks into smaller segments. This is similar to Frederick Taylor's examination in 1911 on improving the process back in his day of having humans shovel Pig Iron. While Taylor looked at chopping up, segmenting, and revising work responsibilities to improve the task process plus working conditions, Ritzer examined the changes plus real-life application of the same concept 80 years after Taylor. Knowing how much I adore Ritzer's works I purchased *The Globalization of Nothing* (2004) and changed my life forever.

The purpose of *Our Six Dimensions* is to tell you how I came about discovering, identifying, and testing our six dimensions plus what they mean to us. The scope of *Our Six Dimensions* is wide plus deep with me bringing you alongside in my data analysis, findings, and applications. I want you to figuratively be next to me so you can envision and walk through my discovery process. This book is written very similar to you sitting with me and having a conversation. I pull in earlier mentions to connect with current discussion.

The discovery of these dimensions is amazing because they topple down prior theories and allow for stronger philosophies to be assessed and implemented. On equal standing with rebuilding stronger theories and philosophies these dimensions provide foundational insights for us to operationalize into everyday life.

This book is not for everyone. Once I noticed the six dimensions then the philosophical question came up where I had to answer 'Do you tell people the truth even though the truth is new and different.' My findings are not what people have been told to believe and the truth may contradict current beliefs. So do you let people live in ignorance or do you tell them the truth. I believe in the truth plus bringing topics to the table to have open conversations about them. Open discussions allows testing to be verified as well as beliefs to be recognized and identified. Exposing people to the truth allows them to know the truth plus gives people the opportunity to choose whether they want to accept the truth. In case some of your knowledge and beliefs merit updating and revision to the extent of causing you to break through your box of limitations, then retaining your logic and objectivity becomes very important. Everything is data and this is new data, so your logic and objectivity really becomes important.

My background is sociology and urban studies and I am a PhD of Urban Studies, so I save society plus cities. I am not an MD who saves lives. Consult a physician before implementing anything.

Recognize all business names within this book are registered trademarks or copyrights, however ™, ®, and ˢᴹ use up space so within this book only business names are listed. Legal protection of business names are acknowledged.

Peer review for my research is strong and very solid, but I decided to not publish my findings in research journals because their 7,000 word limit would be difficult to honor. As well the process of getting these six dimensions into everyone's hands would be long, convoluted, and involve territorial plus ego battles. I am more focused on everyone having access to this knowledge because these dimensions are important.

As well the purpose of this book is to inform and educate, so I reference and quote theories and their theorists. I also argue in support plus against the strengths plus weaknesses of theories. This is because volleying theories allows us to discern insights to create stronger theories. We are all on the same team of examining information, eliciting parsimony, and looking for outliers. I have included three papers and a joke, but cannot find written versions to reference and I apologize. All quotes are under 250 words and single source cumulative quotes do not exceed four in number or over 800 words. Yes, I am a little tight on my wording because each page costs money to print and time to read. In my first proof of this book I listed the six dimensions within the first 63 words and the remaining pages discussed my how's and why's. I was rejected within a couple of hours after submitting that proposal to a literary agent, hence you get the deluxe debriefing.

Do you remember those psychology papers researching problem solving skills of monkeys and the basic concept is the researchers use a monkey, a banana just out of reach, and a box to be used as a tool in the monkey's problem solving. So, the monkey has to recognize its' current knowledge set is limited and not enough to get the reward. Once the monkey releases their limitation of only using past knowledge and the monkey quits blocking their future change then occurs and the monkey is empowered towards expanding their problem-solving skills and getting the banana. At times I felt like that lab monkey because this discovery plus running statistical tests to assess these dimensions has definitely caused me to release prior beliefs and knowledge in order to understand these six dimensions.

There is also a psychology paper using the same premise and the researchers use dogs, dog treats, and maybe a widget to be used to get the reward or a logic path which is intentionally blocked. The researchers concluded since some dogs barked to get their human to help them in problem solving, those dogs were not as intelligent as comparable monkeys who moved the box to get the banana. In generalized terms I see dogs are closer in their relationships with their human care givers compared to lab monkeys, then dogs used their relationship as an extra tool from their problem-solving toolbox. And this extra tool led to asking for help from their humans. From this research gleam the importance of understanding your projects in life, school, work, or research. Learn from people with prior knowledge plus learn from your mistakes. It is OK to test for new insights. If I had not tested these six dimensions, we would still be living in Einsteinian three dimensions.

These six dimensions were observed, tested, and supported by an urban studies/ sociology researcher. Hard sciences had over 2,000 years to find these dimensions and got as far as three dimensions plus no tangible theory of everything or unified field dimensions. Because a soft-science or human-science identified these dimensions the importance of the human perspective cannot be overlooked. Life is not limited to widgets, fulcrums, dials, and formulas. If all one has is quantitative skills then one will view reality from a quantitative perspective and miss out on compassion, diligence, and responsibility as a part of one's lives.

Most published research is statistically significant or substantively significant and this book is both. My test of the six dimensions plus my validity testing to verify my testing are statistically significant. These findings are substantively significant because these findings are our tools to help others within society which in turn helps us as society. I am done with being disappointed in statistically significant research identifying how society is breaking apart from lack of support. Research studies support we as society have performed an excellent job at destroying our weave of social connections by chipping away, nicking at, and slashing out extensive cut-outs to leave our social fiber brittle and frail. In many places we have bitterly and caustically burned away semblances of humanity, people have twisted around and diminished into dark holes of negative existence, and some have devalued all life enough to format pedestals praising Ugly Hatred.

This is our turning point. Use this research as a tool to become the line leader by overcoming obstacles, rising to inspire others, and aiding them. My charge is to inform and explain these dimensions to you plus task you in using this enlightenment. You have the keys to empower humanity and humanity's needs are as diverse as our strengths.

Similar to mitochondria, your specialty knowledge and skills are the energy powerhouse to convert this fresh insight of six dimensions into helping works of the heart. Be the adviser, leader, connecter, and facilitator who strengthens and improves people's lives. Breaking through limitations plus these six dimensions are tools waiting to be used in understanding the universe plus empowering others towards improving themselves to have a better life. Healing society fortifies our social connections through thick robust weaves to create humanity's safety nets of much needed social connections.

In *The Gift* Mauss (1990/ 1950) powerfully unwraps and peels back layers of social exchange and gift giving. His examination elicits the simplest nod or 'hello' as a transaction, the creation of a memory between people, and a gift intent towards bonding with others. In contrast when your ego transacts hatred, unattended rage, bias and anger towards others, your ego gifts them an ugly relationship. Ride herd on your ego because that is the wrong battle and this dysfunction has to change. We need to revise our transactions. Being ugly is a cheap fear-based emotion and gains you no points. Use these dimensions on your hero's journey (Campbell 1949). I mean you no harm and this is my voice and my truth.

Overview

Because I identified our six dimensions when researching how to measure businesses as urban cultural icons this book follows the time sequence of discussing urban cultural icons first, followed by the six dimensions, and then a little braiding of the two topics.

The chapters are as follows:

- Chapter One has an overview, provides direction, and explains the pre-curser and creation of the first five dimensions. I debrief how our six dimensions were discovered while I was researching formatting businesses as urban cultural icons. I also discuss including a spike eliciting the sixth dimension.

  I debrief about the discovery and testing of the five plus one Über-dimension because once the history and naming of these dimensions are discussed, then measuring businesses as urban cultural icons come into play followed by testing the validity of the five dimensions plus transactions.

- Chapter Two establishes my choice of using published research papers as the data set to test our six dimensions. Research findings support the six dimensions 100%. In contrast only 1.1% (four cases of the data set) support Einstein's three dimensions.

- Because I could not find theorists who named and identified dimensions, I switch directions and begin Chapter Three with the topic of measuring businesses as urban cultural icons and proceed into theorists and their concepts of culture and icons since their theories elicited the six dimensions.

- Chapter Four begins with formatting a Guttman scale to measuring businesses by each dimension. For this research neither linear nor logarithmic measurement for Time is the correct measurement, so a modified logarithm is used.

- Chapter Five provides examples to clarify the discussion of each dimension. The role of hypercubes plus schematic framework are included.

- Chapter Six lists and compares Globalization index scores plus Local through Global cultural icon criteria of businesses. Strengths, weaknesses, improvements, and critique of research, plus additional statistics confirm research findings.

- Chapter Seven has us assessing Aristotle's, Albert Ein-stein's, Steven Hawking's, and Kurt Gödel's perspectives against the six dimensions. Three basic linear sequences of the six dimensions are examined. Multiple realities and false dichotomies, some hot spots, and Theory of Everything plus Unified Fields are examined.

Since these dimensions are throughout the universe and our lives, additional stories are included to provide insights and clarify topics. When I identified the six dimensions I could not envision debriefing about updating Albert Einstein's (1961/ 1952, 1920, 1916, 1905) theory of three dimensions. I wrestled with that concern for quite a while. Once I accepted the responsibility, I presented my dual-findings at three conferences: AAG (American Association of Geographers) 2017, SSSA (Southwest Social Science Association) 2017, and UAA (Urban Affairs Association) 2017 and received good reviews. I also submitted my research paper for journal publication and received solid positive reviews, but peer reviewers wanted me to connect to additional topics and expand some discussions. That is impossible to do when looking at 7000 word and a 20 to 30-page limit which includes references.

When it comes to PhD programs some have strict protocols and candidates must research a topic their committee chair assign to them because their committee chair deems this research as worthy. Our program at College of Urban and Public Affairs, University of New Orleans allowed candidates to create their own research. So you research and pick a research topic, create a research question, get this approved and then go about solving for an answer. Since I was in the Urban and Public Affairs program our topics had to be urban-based.

One seminar discussion we volleyed about was 'What is Urban?' and answers ranged from 'Inner-city blacks,' to 'schools and education' and crime. I needed more space so I said 'That which is not rural is urban.'

I wanted to research some form of culture and since I was living in New Orleans I was leaning towards food and how food aids in the creation of culture. I began collecting information on a handful of business case studies from the New Orleans Metropolitan Service Area (MSA).

Starting to Collect Business Cases

K&B Drugs had been a local New Orleans institution for quite a while and the owners were getting old enough to hand over the business to their kids except their kids were not interested in running the business. The owners sold their locations to CVS Pharmacy plus the owners liquidated products and shelving. K&B Drugs even sold off their post-purchase plastic bags imprinted with their signature K&B purple logo. People were buying these things like crazy, people bought K&B private-label match boxes and K&B ball point pens that just the day before were laying on the countertop for use when signing your receipt.

I am not originally from New Orleans, but I recognized K&B Drugs had to be included in my data set. Additional locally-grown businesses included Schwegmann's Grocery, Dorignac's Grocery,

Morning Call Coffee, Bud's Broilers, Café du Monde, Angelo Brocato's Confectionery, plus McKenzie's Bakery and I shopped at all of them. New Orleans also had some non locally-grown businesses to compare against the locals such as Starbucks, Whole Foods, and Baskin Robbins, so the data set was expanding and starting to take shape.

One day I was in Kenner LA dropping off our truck for repairs and I turned to my right to see a big business sign 'Mr. Binky's Adult Shoppe.' The name said it all to me because the name was not steeped in 50-100+ years of New Orleans heritage nor was Mr. Binky's Adult Shoppe a mega international business, so I laughed and thought 'Let's scramble up some eggs and see where this leads. I am adding Mr. Binky's Adult Shoppe to my data set.'

I studied chemistry, physics, and other courses back in the day and a chemistry insight I remembered was to include spikes in one's experiments to test the research method by verifying testing process integrity. Spikes are in surveys and are when you want respondents to mark 'Strongly Disagree' or 'No' to opposing perspectives they have responded 'Yes,' but I have not seen spikes in any other type of research. There is always a first. Mr. Binky's was definitely part of New Orleans metropolitan culture and was definitely in my dataset.

My Research Perspective is Updated

Meanwhile Hurricane Katrina came through New Orleans and

we had reservations and lived at La Quinta in Tomball (near Houston TX) for three to four months. Soon after Hurricane Katrina went through New Orleans a curfew was implemented meaning we were allowed to return home after sunrise and leave before sunset. Our house needed to be gutted plus sprayed with bleach-water to stop mold growth, so we packed food, three gallons of water per person, a full tank of gas plus two full jerry cans, and we were always hungry and exhausted by sunset. Hurricane Katrina destroyed so much of New Orleans that the destruction would be overwhelming if you accepted it personally. We returned to New Orleans on weekends to clean up our house and divide our belongings into two groups: salvage versus pitch. Those were not good times.

Federal funding for universities are based upon student enrollment and presence, so a lot of Louisiana universities were going to be hurting if they could not make classes happen. Since homes, apartments, and UNO buildings were ruined, destroyed, or mangled, professors were provided temporary housing trailers on our campus parking lots and classes were taught under big tents. At this point it was easy for me to figure out my research topic would not be food and culture. The town was so destroyed. Because of this I changed my research focus to measuring businesses as urban cultural icons.

A little bit before Hurricane Katrina occurred a peer asked how I was going to measure businesses to identify them as an icon. Was I going to measure whether businesses have a parking lot, was there a cut-off point of how many parking spots make a business a greater or lesser icon, did the business have electric doors, and what was the number of windows at their locations. Right there I recognized there was a problem.

Windows, doors, and parking spots are superficial and not necessarily indicative of the beliefs, values, and culture of a business. This is because businesses do not necessarily own the location from which they vend plus the needs of businesses change through time. We don't buy horse buggy carts like we used to buy them back in the day. That mis-information of my peer was my benefit because their mis-information aided to springboard me into reverse-engineering Ritzer's continuum from his *Globalization of Nothing* (2004).

We were still living at La Quinta Hotel in Tomball TX when Hurricane Rita came through Houston. It was a bit ironic to watch from our hotel window Houstonites evacuating out of town ahead of Hurricane Rita. We had nowhere to evacuate towards plus our belongings were *safe* in New Orleans, so we might as well stay and keep our hotel room. After Hurricane Rita passed the next door WhataBurger was closed so we walked to a drug store to buy frozen food to live off of until the grocery store opened back up. There were lines in the drug store with people buying disposable cameras, food, and water. All we needed to buy was food. I am at the frozen food section and grabbing four lots of four meals, a young couple to my left were still deciding what to purchase because she was vegan. I had already learned each minute of delay costs you in diminishing selection variety. The weather slowly calmed down and life went on. I used this time to check out Houston-area businesses for my data set plus I was dis-assembling Ritzer's 'Something-Nothing Continuum.' I added Klein's Grocery (Tomball TX), Taco Cabana (TX), Bad Ass Coffee (TX), Conn's Appliances (TX) and WhataBurger (TX) into my data set.

Figuring Out the Puzzle

I had been reading Ritzer's (2004) book and his chart on page 20 intrigued me and I kept going back to stare at the chart because his chart was a puzzle that had to be deconstructed and put back together. There was a hook in his chart and I needed to figuring out the hook.

Continuums show change in two directions compared to scales which show the same changes, but change for scales start at zero and go to plus in one direction and you can have two scales a scale measuring the increase in negative values and a scale measuring the increase in positive values. I only needed the scale to increase in positive values.

In reviewing Ritzer's "Something-Nothing Continuum listing Five Sub-Continuums" listed in *Globalization of Nothing* (ibid.) I thought instead of using his continuum I need to flip the left side of Ritzer's continuum onto the right side of his continuum. This concept is similar to folding a towel in half and using the middle as a crease line with the left and right selvages lining up. Folding over the continuum converted the continuum into a scale.

I wanted to separate the value scale from the topics of Ritzer's five sub-continuums which meant I needed to revise Ritzer's continuum into a two-column chart listing five topic-headers plus a basic zero-to-plus scale. To display overall change Ritzer used 'Something' and 'Nothing' for his continuum. On the sub-continuum level Ritzer displayed change with continua ranging from 'Unique' to 'Generic,' 'Local Geographic Ties' to 'Lack of Local Ties,' 'Specific to the Times' to 'Time-less,' 'Humanized' to 'DeHumanized,' and 'Enchanted' to 'Disenchanted.' See Figure 1.

By flipping over Ritzer's continuum onto itself I converted the overall measurement values into scales with measurement values signifying change by ranging from zero to plus (**0** to +). The sub-continuums in Ritzer's example became my Units Receiving Assigned Value and there were five topics: Entity, Place, Time, Transactions, and Constants respectively. Figure 2 is the rawest form of our dimensions and those are the first five of our dimensions. See Figure 2 for my revisions.

My conversion identified two findings listed in Figure 2 columns: The Units of Analysis which identifies that which is being measured plus a measurement range to quantify their change from nothing to plus. When I put Mr. Binky's against the five dimensions I saw a hidden dimension come out, similar to those 3D art prints where an image visually pops-out at the viewer. This dimension was not overt with any of my other business cases.

Something	Nothing
• Unique (One-of-a-Kind)	• Generic (Interchangable)
• Local Geographic Ties	• Lack of Local Ties
• Specific to the Times	• Time-less
• Humanized	• Dehumanized
• Enchanted	• Disenchanted

Ritzer's value range from Zero (0) to Plus (+)	Unit Receiving Assigned Value
• Unique to Generic	• Entity
• Local Ties to lack of ties	• Place
• Specific to times through no reference to time	• Time
• Humanized to mechanized	• Exchanges/ Transactions
• Patterned similarity to jumbled mess	• Constants

Figures 1 (top figure) and 2 (bottom figure): Figure 1 Source: Ritzer, George "Something-Nothing Continuum listing Five Sub-Continuums," (2004) *The Globalization of Nothing*, p. 20. Figure 2 Source: Author, Author's Translation and Conversion into Units of Analysis Receiving Change.

The stigma of Mr. Binky's Adult Shoppe was enough to elicit

the Human header/ dimension title, expanding my headers/ theme titles to six.

I also noticed a couple of limitations on Ritzer's continuum:

- Ritzer's sub-continuums were independent without interconnection between sub continuums. Ritzer wrote about plates one eats from for one continuum but does not use those same plates when measuring a-nother continuum.

- When I tested Mr. Binky's Adult Shoppe with Ritzer's continuums the lack of the human dimension in Ritzer's continuums was evidenced. This lack was because through Ritzer's lens the adult part of Mr. Binky's was not recognized and without that recog-nition Mr. Binky's would equal other retail businesses such as K&B Drugs or Café du Monde as cultural icons of New Orleans with Mr. Binky's stigma mis-sing. This insight completed the conversion of Ritzer's five sub-continuums into my six dimensions.

Using Mr. Binky's Adult Shoppe as a spike was successful in discovering the omission that was just not overt enough to be observ-ed from other business cases. Mr. Binky's Adult Shoppe elicited Human, the sixth dimension. When I combined my five dimensions from Ritzer's continuum with Mr. Binky's Adult Shoppe and produced Human, that was the first sighting of our six dimensions. The first five of our six dimensions were easier in identifying, but when I was measuring Mr. Binky's Adult Shoppe with this pre-liminary five-dimension scale I noticed Vice (Human dimension) was not evidenced/ showing itself and yet Mr. Binky's Adult Shoppe sold adult intimacy and erotic products. The stigma of Mr. Binky's Adult Shoppe was enough to elicit the Human category.

Our six dimensions are: Time, transaction, entity, place, constants, and human. I saw these six dimension and listed each of them individually in my dissertation (Lawrence 2008) without identifying them as the six dimensions of everything. My dissertation had to focus on urban research, so the six dimensions as their own group would have to wait until after my PhD was awarded.

Scales and Increasing/ Decreasing Values

Since scales and units of analysis carry an important role I want to take a moment to provide some insights on scales and base value and then I will discuss change, theories, and examples. Karl Marx (1992) examined capitalism and using land as an example once there is value-added to the land i.e., homes, rental properties, or business buildings, a simple scale of two data points involving a base value plus an updated value added has been created. From another perspective Ian McHarg (1995/ 1969) also examines scales as elevation scales in *Design With Nature*. All in all, scales, elevation, and capitalism start with a base value and then change to a new value. Adding value creates two data points and a scale. Scales can increase and also decrease.

There is an academic paper from the 60s and I cannot remember the title, but the paper discussed medical students learning how to dis-engage from their emotions and learn how to be objective while performing autopsies. These pre-med students taught themselves to bring a + (high) value event into a 0 (neutral) value event by converting or scaling down the emotional value of the event.

To clear emotional triggers and go from + (high) value negative memories to 0 (neutral) value negative memories and get rid of those dings in life envision a highly negative emotional-value memory and slowly envision physically shrinking and simultaneously bringing the emotional value down to 0.05. If the memory is brought to a zero emotional value there would be no compassion, rather 0.05 is enough compassion to not be a cold-hearted person while enabling the ability to examine, think through, and understand the event better. You just implemented change to decrease the emotional value of an event and it ends up we have change within all six dimensions.

I identified, tested, and reported the six dimensions of everything and a key characteristic of these dimensions I want you to remember was nothing gets activated until change occurs. Because of this look for change within and between these dimensions. A couple of examples of change and dimensions are changes through time which is known as history. Some examples of themed changes through time are medical history, petroleum retrieval history, military science history, transportation history, and the history of inventions. A simple change is humans showing change from their start point of being born at 9 pound 3 ounces and growing from therein. Change can also involve two or more dimensions.

An example of change involving two dimensions is change through time and difference of opinions. This theory is differing age groups of humans having differing beliefs and this is because age groups are sections of time used to categorize humans and humans have various beliefs. If you add another dimension, place, into this example then you have the difference of human beliefs identified by age groups and place of residence. Research wording for that topic is the beliefs of people differ as they age and beliefs differ based upon regionally-based customs. Another example of change is the change of place which could be a city expanding its boundaries and becoming a larger size or a section of town changing its built environment through time via change through demolition, development, and zoning.

Six Dimensions in Real Life

As entities, music and art also have change. One can hear the overtones plus undercurrents of rhythms or patterned change in music. Additionally, one can see written change in music when reading sheet music. The syncopation changes within the "One Note Samba" (Jobim 2008/ 1959) plus introduction of additional musical notes is a great example of change from a singular data point. Jobim uses more than one note after a bit of music, which is change, but listen to the tempo change of the one musical note being played.

Patterned change and difference in kind and degree can also be observed in artwork. Picasso's (1958) "Bouquet of Peace" canvas-art is one of my favorite pieces because there are differences in color plus he drew a transaction of flowers being gifted to another person. When you envision Picasso's (1957) dachshund drawing, compare and then identify changes between his dachshund drawing and "Bouquet of Peace" (1958). These musical and art examples are human-produced change and are not random changes occurring in nature rather these examples allow linear-thinking minds to under-stand that patterned quantifiable change is within the arts.

A hands-on example of entity and other dimensions occurred when I was driving with a buddy along I70 going west from Denver to Grand Junction CO. I was explaining the dimensions and pointing out relatable examples. I pointed out pine and aspen trees, wildlife, other cars and trucks driving I70, and highway signs as tangible ex-amples of entities for my buddy to observe.

I also want you to envision the sheet music for "One Note Samba" (Jobim 1959) as a translucent sheet of the music lines and lay on top of that the sheet music for another tune, and mentally grab the sheet music for your next 1,000 tunes. Now envision all the artwork in the world and add translucent layers of the artwork on top of your music layers. On top of those layers of stacked artwork, stack all the translucent sheets of patterned changes you recognize from reality, also stack all the patterned change observed in all the published papers in all of the academic journals, and add in all the changes in fiction plus non-fiction books. I want you to envision all theories, patterns, art, music and patterned changes recognized from reality. Now look down through your stack of envisioned translucent sheets of patterned change and you will see gallimaufry or a jumbled mess better known as Chaos theory (Gleick 1987).

Everything appears to be random and yet we know layers of patterned change created that chaos because we stacked the translucent sheets. Theories, patterns, and examples will be pulled from this stack of chaos and included in my discussions. Everything is a data point within patterned change, you just have to find the patterned change.

Everything is Constant, But Some Constants Have Names

The conceptual drawing of Figure 3 shows overall everything is a constant and belongs under the grand dimension of Constant. By introducing change then dimensions can be measured and identified. You are at least one minute older compared to when you started reading this book. Look outdoors at the trees and flowers and see how the leaves on plants differ, the cars moving along the street differ by speed, color, or model, maybe their radios were or were not playing, and maybe different radio stations were playing different songs or sportscasts. There were differences between the drivers of those cars, pedestrians traveling from Point A to Point B, bicycles, or the speed of buses compared to the cars.

Look indoors at your furniture, coffee or artwork and start seeing them as a type of entity. Time is around you. Your clock evidences time and your paycheck stub also belongs under the Time heading. When did you wake up today and when is lunch because those times go under the Time heading. As well your number of years of work experience or career-professional experience and your buddy's years of experience go under the heading of time.

For transactions your paychecks show the transaction of your time for money, your coffee cooled down so that is a thermodynamics transaction, you use some form of transaction to get into your office and that is a change-in-place transaction. Did the second hand on your clock just click? That is a transaction identifying the change in time. You sitting on your chair and you thinking about the act of sitting are transactions or conversions of energy. See Figure 3.

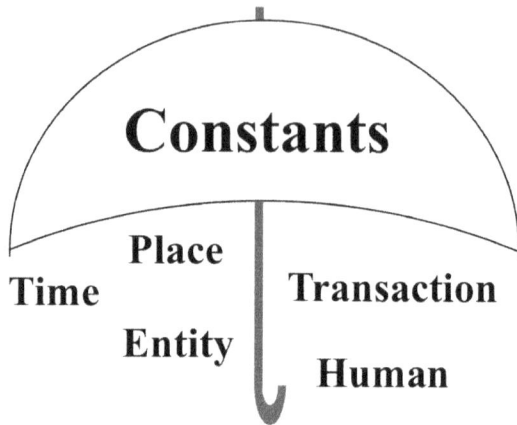

Figure 3 Source: Author, Conceptual Drawing of Dimensions.

Your office furniture and artwork are easier examples to put under the entity header, same with your chalkboard with theories and ideas scribbled on it. Add your sack lunch and your sunglasses, a favorite book and your pet under the entity heading and you should begin to recognize the six dimensions in your daily life. At this point of the conversation all of these are entities, later on I discuss hyper-cubes and constants plus how they interplay with entities.

Leaving the office to discuss various examples of place we have the location of your favorite planet, the area of the Pacific Ocean, the area of a tennis or basketball court, the height of Mt. Shasta, and the flatbed part of tow trucks. The inside area of my crock pot is place and for that matter the bottom of my crock pot is also place. Place is also the distance your children walk from home to school, the number of acres used by a corn maze, where you position the pair of jack-o-lantern's eyes onto the pumpkin or the snowperson's carrot nose plus pair of charcoal eyes.

At this point the six dimensions and the concept of their differences should begin to abound and take life around you and it is startling to recognize these dimensions have always been around us but we never saw them. These six dimensions are ubiquitous and are as germane as cell division or reading this book because cell division is a transaction within an entity and reading this book is a series of transactions, or process, which converts the entity of words into the entity of thoughts within the reader's mind.

I also need you to recognize change within and between the dimensions are a basic part of research papers. Within their specialty topic researchers document and discuss their observations of at least one type of change. There are about 100,000 academic journals available worldwide, so if we estimate four issues per year and ten research papers per journal issue then annually there are at least 4,000,000 documented changes within or between dimensions. Change is written as: Decrease in cellular growth, increase in temperature, relocation of nests, migration of animals, loss of appetite, change in crystalline structure, cancer growth, liver decline, evolution in food storage, preservation, preparation, plus many other examples. Each paper within each issue of a journal has at least one change. As well change is occurring in nature and during regular life.

Right, But How Do You TEST Dimensions?

My dissertation research went well and I was happy with everything in the research. Up to this point the operationalization of my research went well, but my write up of my research was insanely difficult because I needed to cluster and name the collective group of the six header titles that I was using in my study. The problem of figuring out the connective thread which linked time, place, trans-actions, entity, constants, and human together under one umbrella was extremely difficult because this was not the same as listing caring, compassion, and affection and saying 'Oh, that's Love.' There was no name header/ title to put above those six words because no one had grouped them together.

At the same time the initial research purpose of my study was to format a Guttman Scale to quantify (measure) brick and mortar retail businesses as cultural icons and test the scale for reliability (does the scale always measure the same and get logical results) and validity (am I measuring correctly). The key insight to remember is if I had not decided to figure out a way to measure businesses as markers of our lifestyles then I would not have noticed the six dimensions.

During and after this book I want you to meld together these studies, examples, and findings and get past superficial memorization to truly understand my debriefing. I identified the six dimensions dur-ing my research on businesses as urban cultural icons, but the dimen-sions weren't the focus of my study because I was writing my disser-tation. And when one is in the Urban Studies program, then one writes an urban dissertation. The dimensions were observed and identified during my dissertation, then it took a while for me to figure out how to test the dimensions to confirm I was seeing what I was seeing.

When I began to view businesses as markers or identifiers of our everyday lives, then the concept of businesses as icons began to expand. But in academia one cannot just have an idea rather one converts that idea into a research design framework and starts thinking about how to design the research, or in academic terms quantify the variables in order to operationalize the concept.

By the time my dissertation was published and I was awarded my PhD I knew the dimensions were dimensions, but I could not state this knowledge because I had no data to support my findings. My measuring businesses as urban cultural icons are a real-life example of the six dimensions in action. After I defended my dissertation I worked on three projects: Create UBC (Unauthorized Border Crosser) paths in Arizona by using longitude & latitude sites of UBC recorded deaths, expand the research on businesses as urban cultural icons to include more businesses as spikes, plus figure out what type of statistical test to use to observe if these six dimensions really were reality and we have been bamboozled all this time. Businesses as Urban Cultural Icons uses the six dimensions but using them does not support they are the six dimensions. The dimensions have been identified and now we have to figure out how to test their exclusivity.

2

I Used Research Papers as My Data Set

Life settled down after Hurricane Katrina and we relocated to Denver CO on a hillside in the foothills of the Rocky Mountains. I had already lived through earthquakes in Southern California, so we figured there were only a few remaining disasters we had not encountered. Being on a hillside near the Rockies was the most hedge we could get from natural disasters. The hillside home was a safe choice because if flood waters occurred the waters would fill up the Denver basin first. Also living in the foothills of the Rocky Mountains means you are less likely to suffer a tornado because tornados need a bit of distance away from the Rockies in order to form. Ends up hail storms have a couple of favorite terrains and foothills is one of them. Our vegetable garden now has a hail guard made from PVC plus screen door netting. I defended my dissertation and was awarded my PhD in May 2008.

Dimension's Test Takes Shape & More Business Cases Included

One of my dissertation co-chairs, John Wildgen, was living in Arizona and John wanted to perform Arizona-based research since he lived in Arizona half of the year. John was provided locations plus secondary information on the sites in Arizona where UBCs died while crossing the Sonoran Desert. It took us four years to get the paper published and two findings from our paper (Lawrence and Wildgen 2012) were:

- We re-purposed a water flow/ watershed simulation program to format UBC travel paths for crossing the desert. Instead of the program's change in terrain slope to show watershed, our application of the program used the change in death site densities as an elevation and created travel paths.

- There are basic trails used by UBCs to migrate through Arizona. UBC migration is a tough test in one's life.

John died in the later part of 2012. My life settled and I shopped at King Soopers (grocery store) and Wal-Mart, so I put them in queue for inclusion into my business data set. We went on regional trips to Las Vegas, Los Angeles and San Diego, Utah, Oklahoma, Kansas, Wyoming, and back to Texas a couple of times. From these trips there were some businesses I visited and included into my data set: Buffalo Bill's Irma Hotel in Cody WY (named after his daughter), Brookville Hotel in Abilene KS (they moved from Brookville to Abilene as well they have hotel rooms plus serve fried chicken dinners), Bad Ass Coffee TX (their Houston store closed down subsequent to my research), Foss Drugs CO (a general store located in Golden CO), plus Bellagio, Tropicana, and Caesar's Palace (casinos in Las Vegas NV). Back in the day I had eaten at the Brookville Hotel when the hotel was in Brookville KS and have eaten at their new location in Abilene KS and during prior Las Vegas holidays I visited Stardust plus the above three casinos.

We went to Australia in 2013 and 2014. I returned to Denver and for Thanksgiving 2014 I substituted for a buddy and traveled to Antarctica. Let me tell you, geography professors and teachers do not discuss enough about the Drake Passage. Some Antarctica cruises have been cancelled because of hurricanes in Drake Passage and we were crossing the passage in a boat at night time. I wore three types of anti-sea sickness gizmos plus took Dramamine until I had sea legs. At one of our Antarctica landings our passports were stamped with little penguins and we walked into an oasis of a custom-handcrafted

British-made wooden liquor bar, maximum occupancy had to be a-round 25 people. If only I was studying extreme liquor bars of the world because this was and is the southern-most liquor bar in the world.

While in Australia I frequented Brumby's Bakery (Australian bakery chain with a store in Brisbane), Simon's Gourmet Gallery (their inventory included U.S. products), and Cole's Groceries. Cole's Groceries is similar in concept to any regional U.S. grocery chain plus their inventory included some U.S. products. Baskin Robbins was a few doors down from Simon's Gourmet Gallery and Baskin Robbins ice cream in Brisbane tasted the same as U.S. store locations. I thought about including these cases in my data set, but I already had bakery and grocery regional chains in the data set and locally-grown one-location bars. From my New Orleans days Baskin Robbins was already in my data set.

I also knew I could not publish an expanded research on businesses until after I had tested my hypothesis about the six dimen-sions being dimensions. All I could do was call them by their names such as Time, Place, Entity, etc. I needed to be able to cluster together all the individual names and designate their collective group with a topic header. It took a while to figure out the connective thread which links time, place, transactions, entity, constants, and human together under a grand umbrella-type header because no topic header/ title existed since no one had grouped these dimensions together.

For me the six dimensions had a higher priority for discovery compared to revising my businesses as icons research. After I complete the dimensions research then I could publish an expanded version of the businesses as icons because I would have a name for their group and during this extra time I could add more business cases to my businesses as icons data set. Businesses as urban cultural icons begat the task of me first identifying the Kingdom name (as in Kingdom, Phylum, and Family) listed right above these six dimensions.

To proceed I required clarification that the specific word Dimension was used correctly when describing time, place, entity, transaction, constants, and human. It took the longest time to figure out what to use as my data set for testing the six dimensions. I kept pondering Michel Foucault's *Power* (2001) and finally pulled his theory out of the stack of transparencies to take a serious look at Foucault's writing. Foucault wrote power is in everything and I understood that, but I needed that which had the most power for this statistical test.

In addition to needing that which had the most power for statistical testing I wanted written data compared to oral or observational data. Observational data is great for my urban cultural icons research because I would be observing all 58 business cases and converting my observation into scale values. In contrast written data is objective and tangible plus one criterion in scientific research is the ability of other researchers to test one's findings, so I wanted data that would be easily accessible. Oral data has to be transcribed and I did not want peers saying people's beliefs of dimensions interfered with identifying these dimensions, so my thoughts turned to newspaper articles.

Newspaper articles were a potential candidate because microfiche archives allows for accessing back one to two hundred years. Yellow journalism occurred around the 1890s and if I used newspapers I would have to establish a criteria to assess for levels of adulteration and bias. This would be good and bad because the levels of adulteration and bias within Yellow journalism would identify the human dimension. But archival research involves traveling and traveling costs for long distance travels to depositories throughout the United States, printing copies of statistically chosen articles, and potentially being critiqued for using newspapers as a data set, so newspapers were not the right candidate for my data set. I also thought about non-fiction books as a data set candidate, but I would need 350 books in order to generalize findings to the greater population and the cost for 350 books x \$25 would be approximately \$8750 plus tax. My research did not receive any outside funding and my personal budget is tight, so books joined newspapers in my Not-The-Right candidate group. It seemed that everything was Not-The-Right candidate except for one little gem. That gem is published research papers.

The overall concept of using published research papers as a data set is a meta-analysis and usually researchers are examining research results against each other. An example meta-study would examine ketogenic diets, gender, and other variables to assess for overall gist, similarities, and contrasts.

Every published research paper has power within them, so multiple published research papers as my dataset for the six dimensions research creates cumulative power. As well if my hypothesis is supported via published research papers, then my findings are strong enough to withstand peer criticism since researchers had authored and peer reviewed these very same published research papers. This strength plus starting this research with prior knowledge that more than one way to measure time, transactions, entities, constants, human, and especially place exists means the six dimensions could override Albert Einstein's three dimensions (1961/ 1952, 1920, 1916, 1905).

I started thinking about the dimensions and one of the six dimensions has always intrigued me. Transactions are absolutely fascinating. The idea of Transactions as a hierarchical heading and being more than an exchange of cash for coffee occurred when I remembered an organic chemistry schematic of 'Trans-Parents with Mom and Dad at opposite ends.' Ask an organic chemistry person to draw it for you. Your organic chemistry person can also draw a Ferris Wheel just remember the Trans-Parents is what you want to see. 'Trans-' is the connective form for transactions or exchanges. The insight from Trans-Parents is to look for the equals sign (=) because an equals sign is one of many signifiers of a transaction. To a certain extent any relationship between A & B is a transaction. In contrast, exchanges are two directional arrows ($\rightleftarrows$) with one on top of the other, their arrows pointing in opposite directions to symbolize the relationship of both sides in a two-way transactions. This knowledge led me to collapse exchanges under transactions because exchanges, whether equal or unequal are a type of transaction.

Back in the day I studied physics and other courses. I always kept a physics formula ($F = MA$, the transaction of Force equals Mass x Acceleration) in the back of my mind especially when driving on icy roads or seeing fully loaded 18-wheelers coming downhill on I70 a smidge fast. Additionally, throughout my research in sociology and urban studies I would compare conversational transactions, spouse abuse, and the act of teaching against $F = MA$. This is because the equals sign is a transaction moniker, as well $>$, $<$, $\neq$, $\rightarrow$, and $\rightleftarrows$ are symbols for types of transactions.

Then I started thinking since chemistry plus physics, and sociology plus psychology, and urban studies have transactions then transactions occur across academic borders and transactions really could be a dimension. It ends up five of these six dimensions are general enough to occur across academic borders with the sixth dimension being the specialty knowledges occurring within each academic school. Turned another way all of these six dimensions occur across academic borders, but Constants in Physics might not be the same Constants in Sociology. All in all these six dimensions could truly be the all-encompassing all-inclusive dimensions which explain, define, and measure everything everywhere, every time. These could be the unified field dimensions which unite all schools of thought with reality to explain absolutely everything. These six dimensions add structure to the Theory of Everything and this perspective shift is about as powerful as Galileo saying to the church that the Earth rotates around the Sun.

Stratified Proportionate Sampling Using Peer Review-ed Papers

Thinking these six dimensions are the crux of everything doesn't mean they are the crux of everything. I needed to start formatting my data set to run some statistics. Because I chose published research papers in research journals as my data set the statistical test I used to observe the prevalence of these six headers is stratified proportionate sampling. This is because research journals are grouped according to the first character of the journal name making the population count for journal names starting with the letter A unequal to the population count of journal names starting with the letter X.

Regardless of their year of publish these papers examined reality and if these six dimensions exist in reality then I should observe dimensions within my data set. Let me also mention the findings of published research papers tend to have statistical or substantive significance plus the content of papers can vary such that if a paper compared theories then potentially time or place might not be included in the discussion. I did not anticipate observing all of these six names in all published papers, but all published papers should include at least one of these six dimensions.

The test was to locate a paper which failed to support these dimensions by not including at least one dimension. The published journal papers were accessed through University of Colorado Denver, Auraria campus library in 2014 and all journals with titles beginning with an alphabetical letter qualified as strata. Journals starting with '0-9' and journals under the 'Other' groupings were excluded.

These two groups could have been included without impacting any results because each of the groups have low enough journal counts that they would not have qualified for any sampling. There were 149 '0-9' journals and 67 'Other' journals and after calculating their respective proportionate amounts to be sampled neither of these two journal groupings had enough value to merit sampling them. From the calculation perspective this means even if these two groups of journals were included in the data set no samples would have been used from these two groups.

For the stratified part of this research's stratified proportionate sampling there are twenty-six strata which means journals that began with one of the 26 letters of the alphabet are included (93,657 journals total) into the sampling pool. The two highest count strata are journals which began with the letter C (C-journals) and journals which began with the letter A (A-journals). C-journals listed 8,891 journals which began with the letter C and A-journals listed 8,263 journals which began with the letter A. The two lowest count strata were Z-journals (361 journals which began with the letter Z) and X-journals (51 journals which began with the letter X).

For the proportionate part of stratified proportionate sampling, higher count journals are sampled more and lower count journals are sampled less. For example, C-journals were sampled for 33 cases because there were 8,891 C-journals listed and that number of C-journals was divided by the entire population of journals (93,657) to calculate the proportion C-journals occupy of the entire journal population. Calculated out 8,891 C-journals/ 93,657 total journals = 0.0949 or 9.49% of the entire journal population are journals which began with the letter C in their title, C-journals. This calculated percentage is then multiplied by how many cases I wanted in my data set (350 total cases) and the answer is the number of cases I needed to pull from C-journals, so 0.0949 x 350 total cases = 33 cases from C-journals.

The same mathematical calculations are performed for all 26-journal strata and the number of cases the second highest strata-journal needed for my data set is 31 cases. This is because there were 8,263 journals which began with the letter A. The calculation produced 31 as the number of cases to pull from A-journals. It makes sense then that lower count letter-journals are sampled less (Z-journals had 1 case; X-journals were not sampled).

The act of sampling comes from randomly generated numbers identifying every nth case until the proportionate number of journals within each strata is counted. Randomly generated numbers are used again to pick papers from within the journals (tables of randomly generated numbers are usually listed in the back of statistics books). To further establish generalized integrity of the data set, research papers that are selected had to be peer-reviewed, with references available, and written in English. The initial sample size was 350 papers since that allows findings to be generalized to the population however rounding error (i.e., 46.3 cases equaled 46 cases) decreased the sample size to 347 papers with four papers considered for deletion but retained because they did not impact the findings.

Testing Results for Dimensions

Three-hundred and forty-seven (347) papers were reviewed and cases ranged from a minimum of three dimensions (Entity [Tobacco Callus], Transaction [Indole-3 Acetic Acid metabolism and ethylene production], and Human [author] in T. Lee, T. Dumas, and J. Jevnikar [1983] "Comparison of the Effects of Glyphosate and Related Compounds on Indole-3-Acetic Acid Metabolism and Ethylene Production in Tobacco Callus") to five dimensions observed ("Comparative Relationships Among Eye Color, Age, and Sex in Three North American Populations of Cooper's Hawks" [Bielefeldt,

Bozek, Grosshuesch, Murphy, Rosenfield and Rosenfield 2003]). Identifying the dimensions are as follows: I use research authors as the human dimension because they own their interpretation of reality. The topics of examination are Entity, intra- or inter- dimensional change occurs in at least one additional dimension, and other dimensions are self-explanatory. Because publishing journals assign publishing dates those dates are not integral to the bodies of research and are not included as measurement of Time. Historical research, timed-response research, surveys of age groups, are some examples evidencing change over time. Research results are listed in Figure 4. For Figure 4 Entity is abbreviated as E, Human is abbreviated as H, Transaction is abbreviated as Tr, Constant is abbreviated as K, and Time is abbreviated as Ti.

The lowest number of dimensions observed in any paper were three dimensions and three dimensions were observed in 199 cases or 57.3 percent of the total cases. Four dimensions were observed in 130 cases or 37.5 percent of the total cases, and five dimensions were observed in 18 cases or 5.2 percent of the total cases, six dimensions were not observed in any cases.

For cases with three dimensions in their research Entity, Human and Transactions had the largest number of cases, 195 cases or 56.2 percent of the total cases. Entity, Human and Place were listed in 1 case or 0.3 percent of the total cases and Entity, Human and Constants were listed in 3 cases or 0.9 percent of the total cases.

For cases with four dimensions in their research Entity, Human, Constants and Place were listed in three cases, or 0.9 percent of the total cases. Entity, Human, Constants, and Time were listed in two cases, or 0.6 percent of the total cases. Entity, Human, Constants, and Transactions were listed in 33 cases, or 9.6 percent of the total cases. They were the second largest group of cases with four dimensions. See Figure 4.

		Number/ Percentage			Number/ Percentage
Three Dimensions	E, H, and Tr	195/ 56.2%			
	E, H, and Place	1/ 0.3%			
	E, H, and K	3/ 0.9%	199/ 57.3%	L x W x D	4/ 1.1%
Four Dimensions	E, H, K, and Place	3/ 0.9%			
	E, H, K, and Ti	2/ 0.6%			
	E, H, K, and Tr	33/ 9.6%			
	E, H, Place, and Ti	1/ 0.3%			
	E, H, Place, and Tr	18/ 5.2%			
	E, H, Ti, and Tr	73/ 21.0%	130/ 37.5%		
Five Dimensions	E, H, Place, Ti, and Tr	11/ 3.2%			
	E, H, K, Place, and Ti	1/ 0.3%			
	E, H, K, Place and Tr	1/ 0.3%			
	E, H, K, Ti, and Tr	5/ 1.4%	18/ 5.2%		
			347/ 100%		4/ 1.1%

Figure 4 Source: Author, Number and Percentage of Dimensions in Dataset.

Entity, Human, Place, and Transaction were listed in 18 cases or 5.2 percent of the total cases and were the third largest group of cases with four dimensions. Entity, Human, Place and Time were listed in one case, or 0.3 percent of the total cases, and Entity, Human, Time, and Transaction had the largest number of cases and were listed in 73 cases, or 21 percent of the total cases.

For cases with five dimensions in their research: Entity,

Human, Place, Time, and Transactions had the largest number of cases and were listed in 11 of the cases, or 9.2 percent of the total cases. Entity, Human, Constants, Place, and Transactions was listed in one of the cases or 0.3 percent of the total cases, additionally Entity, Human, Constants, Place, and Time was listed in one of the cases or 0.3 percent of the total cases, and Entity, Human, Constants, Time, and Transactions were listed in five of the cases or 1.4 percent of the total cases. The four cases of L x W x D are discussed in Chapter Six.

Because no papers with zero dimensions were observed the analysis of these findings support the existence of these six dimensions. I stare at those words because the enormity of their impact is bigger than an atomic explosion. Time, transaction, entity, place, constants, and human are our six dimensions and they encompass everything, everywhere, everyplace, every time, every person, and every transaction.

3

Let's Talk About Theory of Businesses as Urban Cultural Icons

My research for the six dimensions has no framing theories from academic elders because a dominate belief is we live in three dimensions. None of us questioned Albert Einstein. Some mathematicians have calculated eight and up to 30 dimensions but calculating without naming the dimensions is not fodder for a strong discussion of contributing theories for these dimensions. As well, there has been no research examining these six words, Time, Transaction, Entity,

Place, Constant, and Human as our dimensions.

Academics traditionally have framing theories to use in formatting their research. While there weren't many framing theories naming any of the six dimensions as dimensions I pushed through with testing and identification of the dimensions. In contrast to the limitation of no supporting research to guide my testing of the six dimensions, for my research of businesses as urban cultural icons I had access to many theorists who examined culture, icons, and businesses. Their words set the stage towards formatting the research of businesses as urban cultural icons through using these six dimensions.

Theories Contributing Towards Businesses as Urban Cultural Icons

Businesses exist in our six dimensions and in order to measure them as urban cultural icons we need to look at culture, businesses, and totems or icons. With this we take a break from only discussing the six dimensions to pull contributing theories from our transparencies stack and examine the theory of businesses as urban cultural icons inclusion of dimensions.

Two concepts of cultural turn lead into businesses as urban cultural icons. The first concept for cultural turn is the quantifiable change of culture as in the process of how values or beliefs change from Time A to Time B. This change can be measured through comparison of urban artifacts much the same as archeologists use earthen jugs and relics to characterize early societies. Sanborn Fire Insurance Maps provides an excellent data source to observe cultural change of the first wave of globalization (Ziegelman 2011; Glazer and Moynihan 1970). Throughout the years these fire maps show the turn from tradesman old-world products towards scientific management and economies of scale (Taylor 1911) which is the change in entity and transactions through time.

The second globalization-wave cultural turn shows a growth increase of independent businesses, chain gas stations at street corners, chain restaurants, and retail businesses (Lebhar 1953). The second cultural turn in the United States is two-pronged with one prong occurring after the Vietnam Conflict. The decline of company-owned gas stations and independent businesses were replaced with independent gas convenience stores and big box businesses, an increase in outlet malls with many retailers unable to adapt permanently closing their doors (Basker 2005a, 2005b; Hicks and Wilburn 2001). These specific cultural turns quantify the change of entity and place through time with mention of transactions, constants, and human dimension.

The second definition of cultural turn's other prong is the shift around the 1970s within academia from positivism towards Post-Modernity, post-structuralism, plus ethnic studies. This second definition completes the concept of quantifying businesses as cultural icons because human culture, transactions, and constants were a growing research area during that time and are integral to this research. Clifford Geertz (1973, p. 35) introduces transactions within and between mankind, time and place as contributors to culture when he contrasted Ward's culture "the control of nature by science and art" against *The One Dimensional Man*:

> The trouble with this kind of view… is that the image of a constant human nature independent of time, place, and circumstance, of studies and professions, transient fashions and temporary opinions, may be an illusion, that what man is may be so entangled with where he is, who he is, and what he believes that it is inseparable from them. It is precisely the consideration of such a possibility that led to the rise of the concept of culture and the decline of the uniformitarian view of man.

Separately neither type of cultural turn provides enough dimensions to fully quantify culture, rather the melding of these cultural turns provide six dimension which contributes towards culture and the measurement of culture's change: time, place, transaction, entity, constant, and human. As well this research is not examining an individual's culture, rather theorists recognize the process of culture plus markers within the process having iconic value. Maurice Halbwachs' (1980/ 1950, Pp. 136-7) observation of collective memory through human bonds/ transactions continually weaving through time sets the premise for this study because collective memory allows for businesses to carry quantifiable iconic value using time, place, the businesses themselves, transactions, constants and humans. As well Halbwachs' collective memory pro-vides urban substance to Emile

Durkheim's physical objects being assigned value providing objects with sacred powers via totemism (1995/ 1912) and Marcel Mauss's (1990/ 1950) examination of the economics of social exchanges in *The Gift*.

Sociology v Geography's Icons plus Totemism

Pulling out the transparency sheet of Durkheim's (1995/ 1912) examination of religion and the importance of religious icons in society plus totemism aids this discussion. The word *icons* sends some sociologists into a tizzy because icon is a loaded word for sociologists. In sociology icons and more specifically 'Religious Icons' are the embodiment of the highest of high values assigned by humans towards an entity to identify the most sacred of sacred. This assignment is usually based upon the person's belief system. For one of my dogs her green squeaky toy is her most sacred of sacred while the exalted sacred for the other dog is food, and a comfy bed is the exalted sacred for the third.

In my GIS (Geographic Information Systems) classes I was dumbfounded when geography professors and students used *icon* so casually in their conversations as if they did not understand the almighty importance of that word. I listened in and watched many of their conversations and they would use *icon* when explaining map legends, stop signs, or children crossing signs. At that moment I realized I had been sensitized to *icon* because in normal conversation icon means a symbolic marker to identify something, so railroad crossing signs next to the road are icons identifying train tracks and potential interaction with trains are coming up real soon. Again, that is not what sociologists see. We see personal value to the potential extent of that value being a religious value assigned to an entity.

Icons are not necessarily limited to religious or sacred belief systems, but icons symbolize or are a shorthand marker for something which carries an inherent level of importance. Importance does not necessarily equal sacred, but the two concepts introduce hierarchies and ranking which connects back to starting at a base value and having change from that base value to a new value. In statistics we use rankings to identify groups and rankings range from nominal to ordinal, interval, and ratio.

Disassembling how icons become markers includes a couple of parts: Transactions assigning value or importance towards something, the entity receiving the assigned value, plus the entity assigning the value. I assigned the highest value towards my wedding ring, my husband, and our children plus the act of assigning the highest value makes my wedding ring, husband, and children sacred to me, but not necessarily sacred to you. To you my wedding ring of 39 years is only worth the current price of gold and my husband plus children are of equal value as all people, so assigned values are personal and not necessarily equally shared with others. Not all individuals will assign the same personal value to the same object, but enough individuals can assign personal value of + to an object to then make the object a high-value community object or group totem.

People assign values ranging from 0 to ++ towards anything and everything and they can also update those assigned values. Totems can change over time. For example, the sacred vehicle one enjoyed as a newly-licensed driver over time can decrease in value into a bucket of bolts that needs to be traded-in at the nearby car dealership. Ray Oldenburg's *The Great Good Place* (1999) is a good example from a geographical perspective of people assigning value to places such as cafes, taverns, and hair stylists to identify places/ businesses as a favorite location which people assigned higher emotional value towards places to eat, watch movies, and transact.

Urban Theories Begin Mentioning Dimensions

The overarching concept of Emile Durkheim's totemism is value assignment towards an entity (1995/ 1912) and in combination with culture being a process exhibiting external reflections of values assigned by individuals towards concepts, objects, beliefs, and cultural icons, etc. then one's favorite quill pen can evolve to a favored fountain, or ball-tip pen (Tönnies 1957/ 1887; Arensberg 1955; Glass 1955; Mumford 1938; Tylor 1924/ 1871). To be clear with culture Alfred Radcliff-Brown (1952, 1931) states when it comes to culture individuals as themselves are unimportant; rather people are the tools or facilitator of culture's continuity. This means artifacts, such as businesses, are an integral part in the process of culture to the extent that:

> There are old hotels, dating from the time of stagecoaches, that continue to be used simply because they are in a memorable location. All these routines and remnants from the past require some sort of collective automatism for their explanation, an enduring rigidity in the thought of certain relationships of businessman and customer. These groups adapt slowly, and in many circumstances demonstrate an extraordinary capacity not to adapt. They long ago designed their boundaries and defined their reactions in relation to a specific configuration of the physical environment....To lose their location in the pocket of a certain street, or in the shadow of some wall or church, would be to lose the support of the tradition that recommends them and gives them their unique reason for existence (Halbwachs 1980/ 1950, Pp. 136-7).

What intrigued me with Halbwachs' quote is in 1950 he is writing about old hotels and stagecoaches to the extent that stagecoaches transported probably transported customers to two of my business cases, Antler's Hotel in CO and Irma Hotel in WY. I substituted 'electric car' plus 'chariot' for stagecoaches to observe for variances in Halbwachs' concept to occur. I expanded Halbwachs' idea to include clothes dryer balls, slide rulers, calculators, cell phones, and computers and his wording supports these inventions. For every observable change of culture, there have been quantifiable markers or icons within the process and these changes have occurred through time, businesses, transactions, places, and people (Mellor 1975, p. 278; Arensberg 1955, p. 1143).

Lewis Mumford provided the foundational piece for Arensberg's work by demonstrating that concurrent to each advance of European culture a new form of city was emerging. In her examinations of urban settings Jane Jacobs (1961, p. 58) adds depth to the cultural process with rich examples of theming and social inter-action, she asks, "…why drinking pop on the stoop differs from drinking pop in the game room, and why getting advice from the grocer or the bartender differs from getting advice from either your next-door neighbor…" Jacobs' inclusion of theming was exciting to read because seeing this aided in me expanding my use of the constants dimension into using two constants in measuring businesses as urban cultural icons.

Theming and product line extension are integral in the interactive effects between mass culture, businesses and society (Gramsci 1975/ 1932; Geertz 1973; Adorno and Horkheimer 1972; Lukács 1968/ 1922) to the extent that theming is a basic indicator of a business's culture and is the internal consistency of businesses and co-mingling of their goal with their product lines. Theming, products, and product line extensions provide businesses the opportunity to a-dapt their response towards revised consumer cultural needs plus provides individuals the opportunity to assign a sense of community to businesses as an external reflection of individual cultural values (Hamilton and Chernev 2010; Kotler 2005; Holt 2004; Gobe and Zyman 2001; Pine and Gilmore 1999; McMillan and Chavis 1986; Blau 1964).

Through her studies Ruth Glass affirmed contemporary Western world cities appear as, "a mirror of abstract impersonal forces-of history, class structure and culture" (1955, p. 5) to the extent that businesses are markers of culture. Tuder (2015) wrote, "Think of Taco Bell as a cultural icon … Fast food chains are as much a part of the fabric and history of the United States as any other element."

Two business theories to know are Open Systems and Neo-institutionalism. Through open systems product lines change to reflect change in culture and through neo-institutionalism businesses use dimensions with myth and images to increase their legitimacy in becoming markers of society (Powell and DiMaggio 1991; Katz and Kahn 1966). This means movie theaters, hair salons, adult stores, home improvement stores, hotels, and restaurants are cultural icons or markers of our society and by trading with them we assign value to the businesses and create businesses into sacred markers or totems.

No dimension alone quantifies a business as a cultural icon because time alone does not recognize the geographical expansion of Baskin Robbins stores into Saudi Arabia as markers of culture nor does place recognize theming or time. As well food and supply businesses are basic to cities as are zoning restrictions guiding business placement. The character of businesses merits inclusion because transaction types have different levels of human contact. The theming of businesses pulls from Goffman's *Presentation of Self in Everyday Life* (1959) by applying the concept of how people pre-sent themselves in daily life to how businesses present themselves in daily life. Product line extension is the adaptability of businesses to thrive and/ or push in changing cultures.

Goffman wrote about *Stigma* (1963) which segues well with businesses selling vice. Mr. Binky's Adult Shoppe carries all three types of stigma: Character traits, physical stigma, and group identity. Mr. Binky's Adult Shoppe has "blemishes of individual character" (ibid. p. 7) through selling sex paraphernalia, their physical stigma is having Adult Shoppe in their name plus including Mr. Binky's as part of the business name, plus being a sex paraphernalia shop carries group identity stigma. I visited Fascinations in Lakewood CO and while they retail the same basic sex paraphernalia as Mr. Binky's A-dult Shoppe, the name Fascinations is subtler compared to Mr. Binky's Adult Shoppe which decreases physical stigma and group identity. Overall businesses carry iconic value which can be quan-tified by melding time, place, the businesses themselves, transactions, constants (theming and product line extension classified under the constants heading) and the human dimension.

Businesses Viewed From Urban Perspective

Holt (2004) wrote about the process of businesses becoming icons through reflecting one's cultural values in business's adver-tising and products. I agree with Holt that businesses and brands are icons plus the focus of business is to make a profit. A limitation of

Holt's discussion is businesses using cultural branding to tap into consumer's sense of community or identity guides one-directional transactions towards purchasing that business's brands.

Businesses following this template want to create themselves as icons to leveraging relationships to acquire money. This limitation does not provide an inclusive assessment of the transactional entirety because the sacredness of people, transactions, plus businesses are not acknowledged. Businesses are an integral part of our society's thick weaving of human connections through time, transactions, entities, places, and various constants supported by Halbwachs' collective memory. There is a distinct difference between businesses that have a sense of goodwill and social contribution compared against businesses with a goal of focusing on facilitating relationships to benefit the business.

All in all, Holt provides an excellent example for Tönnies' (1957/ 1887) topic from 100+ years ago *Gemeinschaft und Gesellschaft*. Pulling Tönnies' transparency from the stack we observe both types of transactions plus gradients in between exist. Tönnies observed there are two basic social relationships or transaction types: Gemeinschaft are transactions with a sense of community or social ties within transactions and gesellschaft includes transactions with a primary goal of self-interest especially making money. I envision gemeinschaft as hand knitted or smocked baby outfits or hand quilted blankets gifted to new mothers. For businesses gemeinschaft can include making money as an objective, but the tenure of transactional history plus business longevity in addition to other dimensional contributions transform gemeinschaft into more than an exchange of money for product.

Regardless of adding on or revising the consumer experience (Pine and Gilmore 1999) the downfall of gesellschaft is the lack of community, social connections, and collective memory through tenure, geographical expansion, type of business, transactions, theming, product line extension, plus vice. The measurement and cumulative score of these six dimensions allow us to measure how businesses contribute to societies.

To identify the impact businesses have towards people, the environment proper, as well the built ecosystem, businesses as urban cultural icons must be viewed from the urban perspective. Because gesellschaft does not address these six dimensions as business characteristics then the most local cultural icons of Grand Bazaar in Tehran and Nisiyama Onsen Keiunkan Hotel in Japan would have equal iconic status as Pussycat Theaters, Rick's Cabaret, Starbucks, and Chevrolet.

These theories are strong enough to contend everything can be icons, specifically humans are icons. This insight elicits the need to seriously assess our selves, the sacredness we espouse, plus evaluate our transactions using Tönnies' *Gemeinschaft und Gesellschaft* (1957/ 1887) as to whether they merit revision.

4

Let's Start Measuring Businesses

My dissertation was rock solid, but due diligence in critiquing the dissertation elicited I needed to perform some tweaking to make me happy in refining my concept of businesses as urban cultural icons in order to present the idea to the public. One soft spot of my dissertation was I chose to use natural breaks in dividing my data set to measure time and I discussed this point in my critique. Natural breaks stuck in my mind as the correct measurement to use because all indicators supported using natural breaks, but natural breaks was not the right measurement to use because I felt like I was being corralled into choosing natural breaks and this comes into play with my discussion of measuring time.

What I saw with my dissertation data set was the influx of immigration waves into United States in addition to returning war veterans and both groups were starting businesses to support themselves. I saw these waves of business startups along the timeline and when data elicits a pattern of breaks within the data set, then natural breaks can be used for measurement of that data. The bummer is using natural breaks as a measurement identifier amplifies the impact from these influx waves. So, while my data and findings supported each other I felt I was viewing everything at the microscopic level. The epiphany from that thought was I realized I needed more spikes and more cases to either emphasize the natural breaks or negate those breaks and end that discussion.

I could add U.S. businesses to my data set, but what I needed was businesses prior to U.S. becoming a nation in 1776. Limiting my businesses to the U.S. ignored businesses starting before 1776. By including older businesses which began prior to 1776 I would expand my data set into a macroscopic level and allow me to not focus only on events in the U.S.

I also needed to include into this post dissertation now I have identified them as dimensions expanded data set more cases of businesses with varying levels of vice especially casinos. Mr. Binky's Adult Shoppe was enough to support vice and the human dimension existing, but I needed to measure types of vice. I took advantage of our travels to nearby western states to include Buffalo Bill's Irma Hotel in Cody WY and other businesses mentioned in Chapter Two. I have limited including histories of these businesses because you can find their histories on the internet, annual reports, and newspaper records.

Businesses Added to the Data Set

During the early infancy of this research I leaned towards examining food and culture, so there are more food-based businesses in the data set especially New Orleans-based food businesses since we were living there. My research began in New Orleans then Hurricane Katrina occurred, and we lived in evacuation housing for a while and then relocated to Denver CO. This means most of the 55 United States businesses are going to have New Orleans, Tomball TX, or Denver locations. Polk Business Directory was the first resource used in counting locations for multi-location businesses, otherwise telephone calls, annual reports, business websites, and news articles were used to triangulate the number of business locations. My limited cash flow in combination with me wanting to observe business transactions limited the data set to 55 cases plus three cases accessed through the internet for a total of 58 cases.

Since I was living in Denver, I included King Soopers (a Denver based, Kroger owned grocery chain) and Albertson's grocery.

A Little Off The Top Hair Salon (CO) was in the Denver metro area and this hair salon had a little bit of vice so I included them into the data set. Since the businesses in my dissertation were U.S. based, so I expanded my new data set to 58 cases by including three pre-1776 businesses from overseas. I chose the oldest business continuing to be operational, the oldest hotel in Japan, and the oldest tavern in Great Britain. These businesses are: Grand Bazaar in Tehran (established around 1600 BCE), Nisiyama Onsen Keiunkan Hotel in Japan (established in 705 AD), and Bingley Arms tavern in Great Britain (established in 953 AD).

Overall Scale Formatting for This Research

As I focused on how to quantify the dimensional value of each business their names became less important as their characteristics increased in importance. This research measures business using all dimensions with businesses as entity, years in operation as time, geographical distribution of stores as place, type of transaction as transaction, two constants (theming and product line extension), and vice as the human dimension. While each dimensional measurement will have differing amounts of categories the assigned value to the start category for each dimensional measurement begins with 0 (zero).

Entity is measured in four categories (food, shelter and clothing, health and personal care and essential furnishings, and transportation with communication and entertainment) with categorical values ranging from 0 (zero) to three. Place is measured in four primary categories, local, focal, zonal, and global, and those primary categories are subdivided with Local having two categories ($Local_0$ and $Local_1$), Focal has four categories ($Focal_0$, $Focal_1$, $Focal_2$, and $Focal_3$), Zonal has five categories ($Zonal_0$, $Zonal_1$, $Zonal_2$, $Zonal_3$, and $Zonal_4$), with Global having five categories ($Global_0$, $Global_1$, $Global_2$, $Global_3$, and $Global_4$). Grand total there are 16 categories for Place with categorical values ranges from 0 (zero) to 15. Time is measured in eight categories (Protype, Archetype, Intertype, Mesotype, Metatype, Supratype, Ultratype, and Pantype) with categorical values ranging from 0 (zero) to seven. Transactions are measured in three categories (Verbal, Countertop, and Conveyor) with categorical values ranging from 0 (zero) to two. Theming is measured in four categories (Minimal, Blended, Fused, and Disjointed) with categorical values ranging from 0 (zero) to three. Product Line extension is measured in five categories (Minor, Line, Low-cost, High-cost, and Heavy Line-replacement) with categorical values ranging from 0 (zero) to four, and Vice is measured in five categories (Outside, In-Store, Hybrid-Innuendo, Multi-Experience, and Sensory) with categorical values ranging from 0 (zero) to six. Each of the 58 businesses are measured in these seven dimensions (five dimensions plus two constants) with their cumulative scores indicating their level of globalization.

Because this study compares between business categories and

not relative price of goods Veblen's (2011/ 1899) conspicuous consumption is not applicable. I use Chris Sarlo's (2006a, 2006b, 2001, 1992) needs-based consumption as the structure to assess and rank the data set of businesses into categories to measure entity within this research. Sarlo's (ibid.) hierarchy ranks needs into nine categories: Food, shelter and clothing, health and personal care, essential furnishings, transportation, communication, and entertainment.

Food-based businesses are ranked in the first category and are assigned the lowest value of 0 (zero) in this study. Businesses which supply shelter and clothing are ranked in the second category and are assigned the value of 1 (one) in this study. Businesses which provide health care, personal care, and essential furnishings (tools, technology, and major appliances) are ranked in the third category and are assigned the value of 2 (two) and businesses which provide transporttation, communication and entertainment are ranked in the fourth category and are assigned the value of 3 (three).

If businesses straddled two categories then their ranking is based upon their primary revenue stream, so Brookville Hotel and Irma Hotels are categorized for their restaurants. The Brookville Hotel in Abilene KS has a restaurant plus hotel rooms and Buffalo Bill built the Irma Hotel with bar, restaurant, and hotel rooms in Cody WY for his daughter Irma. Bellagio, Caesars Palace, Stardust, and Tropicana are categorized as entertainment because of gaming and entertainers performed at their locations. Those casinos straddle two transaction categories (poker tables and live entertainment versus slot machines) with casinos categorized in the lower ranking value of verbal transactions because poker and live entertainment are verbal exchanges. Because these casinos expanded their souvenir selection they sold beyond selling only low-cost souvenirs to also include selling high-cost souvenirs, these casinos are ranked high-cost souvenir on product line extension.

My initial spike was Mr. Binky's Adult Shoppe and I added three additional overseas businesses as spikes. These three spikes included in this expanded data set are:

- The oldest recorded viable business is the Grand Bazaar in Tehran.

- Nisiyama Onsen Keiunkan Hotel in Japan.

- Bingley Arms tavern in Great Britain.

If name extensions occurred (Wal-Mart, Super Wal-Marts, and Wal-Mart Neighborhood Markets) then all similar-named businesses are ranked together (Wal-Mart stores are ranked as variety or dry goods retailers). I thought about a control group of businesses for the data set but realized once a business is operational then the business exists and is measurable in all six dimensions.

Starting the Guttman Scale by Measuring Entity

The first Entity category is food and in New Orleans I shopped at Angelo Brocato's Confectionery, Baskin Robbins, Bud's Broilers, Café du Monde, Dorignac's Grocery, McKenzie's Bakery, Morning Call Coffee, Schwegmann's Grocery, Taco Bell, and Whole Foods. Tomball/ Houston TX food-based businesses I traded at included Bad Ass Coffee (to compare against Café du Monde and Morning Call Coffee), Klein's Market (to compare against Dorignac's, Schwegmann's and Whole Foods), Taco Cabana, and WhataBurger (to compare against Bud's Broilers).

I view Taco Bell as a tangent comparison to burger stands and because Taco Cabana sells beer, Taco Cabana was included to contrast against Taco Bell. Baskin Robbins was visited and included to counter-balance Angelo Brocato's Confectionery. Because Ritzer's book "The McDonaldization of America" (1993) examines the introduction and application of standardization, production lines, and portion control into industry venues, I ate at McDonald's in Denver. I ate at Burger King and Sonic Drive In to contrast against McDonald's which pretty much completes my burger-joint comparisons.

Cheever's Café (was a florist shop which the new owners converted into a restaurant) is in Oklahoma City and Cheever's Café is included to contrast against Applebee's Restaurant in Denver. Starbucks is a must-have since I wanted a wide ranging of cases from single-location to multi-national location businesses. Additionally, Starbucks contrasts against Café du Monde and other coffee vendors. I ate at Brookville Hotel way back when the hotel was located in

Brookville KS plus I ate at its new location in Abilene KS, I also ate at the Irma Hotel in Cody WY. I shopped at King Soopers Grocery in Golden CO and Albertson's in Lakewood CO plus Rock Springs WY, so they were included for their ages and numbers of locations.

Ranking of Entity Dimension	Categories of Businesses
First	**Food:** Basic need for survival. These business types consist of grocery, bakeries, and restaurants.
Second	**Shelter and Clothing:** These businesses provide housing, hotels, and clothing.
Third	**Health Care, Personal Care, and Essential Furnishings:** These businesses include dry goods, drug stores, variety stores, tool and hardware stores, and appliance stores.
Fourth	**Transportation, Communication, and Entertainment:** These businesses include car dealerships, gas stations, phone companies, movies, and liquor-bars.

Figure 5 Source: Author, Entity Ranking and Type of Business.

For the second measurement category of shelter and clothing I used La Quinta Inns and Suites as a case in my study since we evacuated from Hurricane Katrina to La Quinta Inns and Suites in Tomball TX. See Figure 5. I have also stayed at Marriott hotels so they were included in the data set.

The history of Antlers Hotel was interesting enough for me to include Antlers Hotel in the data set. I drove to Colorado Springs CO to walk around and view the hotel. When it comes to clothing stores I did not want to include Dillard's Department Stores since I shop there and I did not want to bias the study by including too many businesses I traded with regularly. Including King Soopers and Wal-mart was good enough for me.

It ends up Dillard's geographical distribution as a southern retail business was important enough for me to override omitting Dillard's, so I included Dillard's into the data set. I also shop at JC Penny but I absolutely wanted to limit including businesses I personally trade at to only one business per category, so nix including JC Penny and keep Dillard's in the data set. F.M. Light and Sons in Steamboat Springs CO sells western wear clothing, so I included them to contrast Dillard's.

I did not think this research was going to work. So when I was deciding Yes to Dillard's and No to JC Penny I was filled with anxiety, self-doubt, second-guessing, and over-thinking. This was when reality set in that my data set was real plus businesses can be quantified worldwide as urban cultural icons.

The third category under Entity is health care, personal care, and essential furnishings and I visited K & B Drugs and CVS Pharmacy in New Orleans LA, Foss Drugs and Walgreen's in Golden CO. For personal care A Little Off The Top Men's Hair Salon had a couple of stores in the Denver Metro area and they were a hair cutting salon where female stylists wore bustier tops and to contrast them I visited

Super Cuts in Denver CO. I traded at Wal-Mart in New Orleans and Lakewood CO. For essential furnishings I visited ApplianceWorld in Denver, Best Buy in Lakewood, Conn's Appliances in Houston, and Home Depot in New Orleans and Golden.

For the fourth category of transportation, communication, and entertainment, I visited Chevrolet in New Orleans, Houston, and Denver. We also owned a Ford and I visited their dealerships but I was beginning to recognize the data set needed additional transportation companies which profoundly varied on the other dimensions in order to have an interesting comparison and robust data set, so I did not include any other dealerships. I looked up the first wheel and axle and they were invented around 5,500 years ago. Chariots existed around 2000 BCE and would have been a great case to include in this dataset, but there are no written records of chariot businesses and there are no written records of chariot businesses currently selling chariots. For communication I did not want to list my telecommunication company, so I visited Sprint in Denver.

For entertainment businesses I visited Applejack Wine & Spirits in Applewood CO and read about Arap's Old Gun Shop in Denver. Arap's Old Gun Shop started as a gun shop but Arap died, so his family took over the business and changed the gun shop to a bar. The bar was open for a while and then closed down. I never visited Mr. Binky's Adult Shoppe in New Orleans because of us evacuating for Hurricane Katrina and then moving west to Denver, so a cohort at

University of New Orleans was generous enough to visit the store for me. I did visit Fascinations in Lakewood and the business concept was the same as the cohort described for Mr. Binky's.

I grew up in Los Angeles CA which means including Pussycat Theaters was a given once I identified alcohol, tobacco, gaming, and nudity as vices legally sold by businesses. To counterbalance Pussycat Theaters, I visited Cinemark Theaters and included them in the data set. Rick's Cabaret is in Houston and I did not visit the business. I visited Tropicana, Stardust, Caesar's Palace, and Bellagio casinos in Las Vegas and those casinos were entertainment complexes compared to selling lodging or selling food at a lodging site. With entity ranking completed I started ranking businesses by their geographical expansion.

Measuring Place for Urban Cultural Icons

The next dimension measured was Place and I combined Michael DeMers (2002) plus Chor Lo and Albert Yeung's (2002) geographical terms with the concept of cellular division/ growth (1, 2, 4, 8, 16, and 32) to measure the geographical distribution of businesses. Their research established using geographical terms of local, focal, zonal, and global in map algebra of geographical data analysis by laying a grid over a study area. The grids then allow for calculations to be performed with cell areas with values assigned to cell areas based upon the majority of what occupied the cell. For example, envision putting a grid over a park and if over half of a cell was flowers, then the cell was coded for flowers and a nearby flowing river would be coded as an inter-connected squiggly-line series of cells. This concept is similar to older computer games where the pictures were blocky or pixelated.

With that idea in mind now let's go back to this geographical definition. The original definition of Local is a basic single cell or square from grid paper. Focal is the eight cell perimeter around the one singular cell, zonal is a bigger area usually with a flowing design such as a wild flower field, lake, or forest, and global is all cells. For this study both the number of stores plus the expansion area covered by the stores are blended to measure their geographical density. That measurement is the premise for the scale categories of place in this study. See Figure 6.

Local has two categories: $Local_0$ is a single store which has always been at the same location and $Local_1$ is a single store that has moved since opening for business. Example businesses of $Local_0$ are Antlers Hotel, Arap's Old Gun Shop, Bellagio, Bingley Arms in Great Britain, Caesars Palace, Cheever's Café, F.M. Light and Sons, Foss Drugs (they remained at their original location and expanded to include retail space next door), the Grand Bazaar in Tehran, Irma Hotel, Judice Inn Burgers, Nisiyama Onsen Keiunkan Hotel in Japan, Stardust, and Tropicana. $Local_1$ examples include AppleJack Wine & Spirits, Angelo Brocato's Confectionery, Brookville Inn, Dorignac's

Grocery, Klein's Grocery, and Morning Call Coffee.

Ranking of Geographical Dimension	Application of Geographical Term towards Business
Local	Local is a single store at a single location. • $Local_0$ is a single store which has always been at the same location. • $Local_1$ is a single store which has moved since opening for business.
Focal	Focal is defined as more than one store and the stores are located within city limits or MSA. • $Focal_0$ has from two stores and up to eight stores with all stores located in the same city. • $Focal_1$ has from two stores and up to eight stores with all stores located within the same Metropolitan Service Area (MSA). • $Focal_2$ has from nine stores and up to 16 stores with all stores located within the same MSA. • $Focal_3$ is businesses with 17+ stores in one state.
Zonal	Zonal covers an area larger than MSA and at maximum the area is contained within one country. • $Zonal_0$ is businesses with locations in two states. • $Zonal_1$ is businesses with store locations in three to eight states. • $Zonal_2$ is businesses in at least nine states, up to 16 states. • $Zonal_3$ is businesses in 17 to 32 states. • $Zonal_4$ is businesses in 33 to 64 states including District of Colombia, U.S. Virgin Islands, Puerto Rico and Guam.
Global	Global is inclusive of all countries, for this study there are five gradients of Global. • $Global_0$ are businesses in at least two countries, up to eight countries. • $Global_1$ are businesses in at least nine countries, up to 16 countries. • $Global_2$ are businesses in 17 to 32 countries. • $Global_3$ are businesses in 33 to 64 countries. • $Global_4$ are businesses in 65+ countries.

Figure 6 Source: Author, Geographical Terminology

and Application.

Focal businesses have from two to 17+ stores with all stores located either within the same city or at maximum, located within the same Metropolitan Service Area (MSA). Zonal-level businesses have state-level expansion and $Zonal_0$ businesses have locations in two states, $Zonal_1$ businesses are in three to eight states, $Zonal_2$ businesses are in nine to 16 states, $Zonal_3$ coded businesses are in 17 to 32 states, and $Zonal_4$ coded businesses are in 33 to 64 states including District of Columbia, U.S. Virgin Islands, Puerto Rico, and Guam.

Global businesses have locations in at least two countries with $Global_0$ businesses in two to eight countries, $Global_1$ have businesses in nine to 16 countries, and $Global_2$ have businesses in 17 to 32 countries. To be clear the website for Chevrolet lists locations in 140+ countries so the listing of Chevrolet as global is based upon their listing of their locations and is not listed by how many stores a dealership owns in the individual dealership's name. Time was the next dimension to measure.

Scaling of Time for Urban Cultural Icons

Books have been written about time, papers discuss bending or slowing of time or time management, but for the most part there are no true Time departments in academia. In my dissertation I used natural breaks for my time categories but the fault and strength in natural breaks is based upon one's data set and that data set was U.S. based businesses. So that data set has inherent skews to the extent that natural breaks enhanced the inherent booms of business start-ups from our immigration/ globalization waves and post-WWII business start-ups which meant the data confirmed U.S. history but failed to identify the absence of pre-1776 businesses.

My initial spike was Mr. Binky's Adult Shoppe, my next spikes were those three business cases from overseas. I figured the oldest recorded business in history would be the best case to include and I included the hotel and tavern for variety and international sake. I limited the overseas spikes to three because I was not able to visit those locations and prove their existence. While these three businesses seem logical and passed the critical eye/ does it walk and quack like a duck/ sniff test, they carried an inherent sketchy factor of -0.5 point. While including 300 businesses from overseas would bring my data set up to 358 cases and provide great cases from older time periods plus make the research robust, I could not add 300 businesses from overseas with my only proof of their existence being their listing on the internet.

Both linear and logarithmic measurements of time performed poorly when used in combination with the other dimensions into formatting a cumulate sum, so the age-grouping of businesses was based upon a modified logarithm of eight categories: Protype (0 to 25 years old), Archetype (26 to 50 years old), Intertype (51 to 100 years old),

Mesotype (101 to 250 years old), Metatype (251 to 500 years old), Supratype (501 to 1,000 years old), Ultratype (1,001 to 2,500 years old), and Pantype (2,501 to 5,000 years old). When critiquing the Time dimension older businesses were greater cultural markers compared to newer businesses. Since the oldest recorded civilization occurred in Mesopotamia around 3300 BCE and the Grand Bazaar began around 1600 BCE (approx. 3,600 years old and listed in the eighth category which ranges from 2,501 to 5,000 years of operation) there is no need for a ninth category.

While many of the cases are businesses which are still open and trading, some businesses which closed are also included in the data set to elicit clarification between the age of a business versus when the business existed. Figure 7 is based upon the age of the business so using K & B Drugs as an example, K & B Drugs opened in 1905 and closed in 1997, meaning K & B Drugs was operational for 92 years and is an Intertype business. See Figure 7.

The first three Time categories cover 0 (zero) to 100 years. Protype (0 to 25 years old) businesses are the youngest of all businesses with little myth, image value or heritage connected to them. Example protype businesses include: A Little Off The Top, ApplianceWorld, Bad Ass Coffee, Cheever's Café, Mr. Binky's Adult Shoppe, and Bellagio.

Archetype (26 to 50 years old) businesses are older than protype businesses and are in the beginning stage of being established.

Example archetype businesses consist of Applebee's, Best Buy, Caesars Palace, Cinemark Theaters, CVS Pharmacy, Home Depot, King Sooper's Grocery, La Quinta Inns, PJ's Coffee, Pussycat Theaters, Rick's Cabaret, Starbucks, Stardust, Sprint, Super Cuts, Taco Cabana, and Whole Foods.

Intertype (51 to 100 years old) businesses are more established plus they are lightly recognized as an institution. Example businesses are Albertson's, Applejack Wine & Spirits, Arap's Old Gun Shop, Baskin Robbins, Brookville Hotel, Bud's Broiler, Burger King, Dillard's Department Store, Dorignac's Grocery, Foss Drugs, Judice Inn, K & B Drugs, Klein's Grocery, Marriott Hotels, McDonald's, McKenzie's Bakery, Mello Joy, Sonic Drive In, Taco Bell, Tropicana, Wal-Mart, and WhataBurger.

The next three time-based rankings begin at 101 years old and end at 1000 years in business. Mesotype (101 to 250 years old) businesses occurred after U.S. was established. These businesses were beginning to become solid institutions with example businesses being Antler's Hotel, Angelo Brocato's Confectionery, Café du Monde, Chevrolet, Conn's Appliances, F.M. Light and Sons, Irma Hotel, Morning Call, Schwegmann's Grocery, and Walgreen's.

Metatype (251 to 500 years old) businesses are in the solid phase of becoming institutions and at this point all businesses from the U.S. do not qualify because the earliest they could be established is 1776. There are no cases in this or Supratype category because the three international businesses I included are the oldest operational tavern, hotel, and bazaar.

Ranking of Time	Phase of Business
0 to 25 years old	**Protype Business:** Business has opened and has zero or very little tradition, myth, image value, or heritage.
26 to 50	**Archetype:** Businesses are becoming more established and are increased in myths, customs, and the like. These businesses are starting to be accepted as common knowledge, having incrementally gained authenticity.
51 to 100	**Intertype:** Businesses are beginning to become institutions. They have been lightly established as an institution.
101 to 250	**Mesotype:** Businesses are in the middle phase of being established as institutions. They are beginning to become solid institutions.
251 to 500	**Metatype:** Businesses are in the solid phase of becoming established as institutions.
501 to 1000	**Supratype:** These businesses are established institutions of industry, but are simple young-established institutions.
1001 to 2500	**Ultratype:** These businesses are beyond simple institutions and are median-established institutions.
2501 to 5000	**Pantype:** These businesses are absolute solid-established institutions of industry.

Figure 7 Source: Author, Time-based Business
Ranking and Phases.

Supratype (501 to 1000 years old) businesses are established businesses and are at the simple established phase. There are no cases in this category.

The last two time-based rankings begin at 1001-year-old businesses and end at 5000-year-old businesses. The three international businesses included in the data set are old enough to be listed in these last two time-based rankings. Pantype (2501 to 5000 years old) businesses are absolute solid established institutes of industry and include the Grand Bazaar established around 1600 BCE.

Business Transaction Categories

The measurement of transactions is a little different compared to the measurement of time. Transaction is the conversion of energy, or the relationship between A and B so in this study Transactions are ranked by the amount of energy used by employees interfacing with customers because the more energy an employee uses in the interface-transaction, the more likely a business is committed to personalizing their service through critical thinking, decision making, solution resolution, and ownership in project management. The three categories for transaction are: Verbal exchange, countertop exchange, and conveyor belt.

In this study Transactions are ranked by the amount of energy used by employees because the more energy an employee uses in the transaction, the more contemporary businesses resemble businesses from one to two hundred years ago. Personalized service does not mean saying a customer's name or saying, 'Have a Nice Day.' Personalized service is the action of catering to, and treating the customer as an honored guest having individualized requirements for their purchase to the extent that labor is performed for that customer and not usually identically repeated for all subsequent customers (five customers in a row could order a vanilla scoop of ice cream but each scoop is their customized order).

Ranking of Transaction	Amount of Energy used by Employee
First	**Verbal exchange:** Employee uses most amount of energy by translating order into finished transaction and purchase.
Second	**Countertop exchange:** Employee uses less energy than verbal exchange; product is usually not individually-made. Purchase is main objective, some verbal exchange.
Third	**Conveyor belt:** Employee uses least amount of energy, customer uses most energy to bring product to exchange. Focus is expedition of transaction.

Figure 8 Source: Author, Ranking of Transaction by Amount of Energy used by Employee.

Labor is a critical cost-marker within businesses so verbal exchanges which convert a customer order into a finished transaction used the most amount of customer-interface based employee energy. Businesses want to be cost effective on paying for labor so in the dyad between customer and employee, verbal-exchange businesses cost more (labor energy plus critical thinking skills) compared against the dyad where the customer figures out and picks out what they want to purchase. See Figure 8.

Example verbal-exchange businesses include: A Little Off The Top, Antlers Hotel, Applebee's, ApplianceWorld, Arap's Old Gun Shop, Bad Ass Coffee, Baskin Robbins, Bellagio, Bingley Arms, Angelo Brocato's Confectionery, Brookville Hotel, Bud's Broiler, Burger King, Café du Monde, Caesars Palace, Cheever's Café, Chevrolet, Cinemark Theaters, F.M. Light and Sons, Grand Bazaar, Judice Inn, Irma Hotel, La Quinta Inns, Marriott Hotels, McDonald's, Mello Joy, Morning Call, Nisiyama Onsen Keiunkan Hotel, PJ's Coffee, Pussycat Theaters, Rick's Cabaret, Sonic Drive In, Sprint, Starbucks, Stardust, Super Cuts, Taco Bell, Taco Cabana, Tropicana, and What-aBurger.

Countertop exchanges have the transaction usually occurring at a countertop or desk and could include some verbal transactions such as 'I'll take a pack of Marlboro.' Less employee energy is used compared to individualized orders (i.e., having the employee roll one's cigarettes for a customer's order would be a verbal exchange, but having the employee grab the pack of standardized pre-packaged cigarettes is a countertop exchange). Example businesses include: Best Buy, Conn's Appliances, CVS Pharmacy, Dillard's Department Store, Foss Drugs, K & B Drugs, Home Depot, McKenzie's Bakery, Mr. Binky's Adult Shoppe, and Walgreens.

For Conveyor belt exchanges the objective of these transactions is to have employee's energy expenditure as close to zero as possible. A conveyor belt is used as part of the transaction process and example businesses include Albertson's, Applejack Wine & Spirits, Dorignac's Grocery, King Sooper's Grocery, Klein's Grocery, Schwegmann's Grocery, Wal-Mart, and Whole Foods.

Theming and Product Line Extension Categories

Two other internal business characteristics which contribute towards their iconic values are theming and product line extensions. The two Constants included in this research are Theming and Product Line extensions and because businesses from one to two hundred years ago have minimal theming, then minimal theming is a characteristic of newer cultural icons since disjointed theming did not exist a couple of hundred years ago.

These constants, Theming and Product Line extensions, are part of the business's personality and while western theming, coffee house chic, and peek-a-boo costumes are conceptually equivalent to Goffman's (1956) "Presentation of Self in Everyday Life" these themes answer the business question of while one is selling the product, how much is one purchasing an experience in the transaction.

The four categories of theming are: Minimal, blended, fused, and dis-jointed. Minimal-theming businesses focused on selling their product with zero to subtle theming or store concept branding. Example businesses are Albertson's, Applejack Wine & Spirits, Angelo Brocato's Confectionery, Café du Monde, CVS Pharmacy,

Dorignac's Grocery, Grand Bazaar, F.M. Light and Sons, Foss Drugs, Judice Inn, K & B Drugs, King Sooper's Grocery, Klein's Grocery, Home Depot, McKenzie's Bakery, Morning Call, Mr. Binky's Adult Shoppe, Schwegmann's Grocery, Wal-Mart, and WhataBurger. See Figure 9.

Blended theming occurs to the extent that the sale of the product is facilitated through the business concept. Some décor theming occurs, but theming does not over-ride the product. Example businesses are A Little Off The Top, Antlers Hotel, ApplianceWorld, Arap's Old Gun Shop, Baskin Robbins, Best Buy, Bingley Arms, Brookville Hotel, Bud's Broiler, Burger King, Cheever's Café, Mc-Donald's, Mello Joy, Nisiyama Onsen Keiunkan Hotel, PJ's Coffee, Sonic Drive In, Super Cuts, Taco Bell, Taco Cabana, and Walgreen's.

Fused Theming is a joined focus on product sale plus theming to the extent that product, décor, and overall concept are premised upon a design. These businesses display focused integration of product, packaging, and concept mixture. Example businesses are Applebee's, Bellagio, Caesars Palace, Chevrolet, Cinemark Theaters, Dillard's Department Store, Irma Hotel, La Quinta Inns, Marriott Hotels, Pussycat Theaters, Rick's Cabaret, Sprint, Starbucks, Stardust, Tropicana, and Whole Foods.

Disjointed Focus is the fourth group and the business concept plus product line do not connect well. Example businesses are Bad Ass Coffee and Conn's Appliances. The Houston location of Bad Ass

Coffee had a 3 ½ ft. tall Plexiglas volcano in their lobby area (that store has since closed) and the Conn's Appliances store sold bed mat--tresses plus high technology equipment.

Ranking of Theming Dimension	Theming Cohesion between Business and Product
First	**Minimal Theming:** Focus is on sale of product with zero to light-level of theming connection between products and ambiance. There is some décor theming, to the extent that the sale of product is facilitated through the business concept, but theming does not over-ride the product.
Second	**Blended Theming:** Focus is on light to medium theming of business blends with sale of product.
Third	**Fused Theming:** A joined focus on sale of the product plus theming to the point that product, décor, and overall concept are premised upon a concept design. These businesses display a focused integration of product, packaging, and concept mixture.
Fourth	**Disjointed Theming:** Fragmented or unconnected linkage between product and ambiance. Disjointed Focus, where the business concept plus product line connect poorly to the extent that concept competes with sale of product.

Figure 9 Source: Author, Ranking of Theming.

The next constant is Product Line extensions. Product Line extensions usually occur with businesses of yesteryear to the extent that if one was selling vinegar, it was not a far reach to also sell cooking oil. An example of product line extension businesses are businesses which modify their product lines to include organic produce lines. Low-cost souvenir business sell their main product plus disposable mementos such as shot glasses, coffee mugs, or t-shirts.

High-cost souvenir businesses have pricier souvenirs such as refurbished old gas station pumps, pinball machines, or rear-ends of old cars converted into love seats. Heavy line-replacement businesses carry the latest technology, have minimal vestige in the past both in their product and store fixtures and are continuously turning over their product lines to carry the newest models of the most recent technology. Advents in technology compels heavy product-line replacement with televisions, telephones, music, refrigerators, and automobiles.

Ranking of Product Line Extension Dimension	Level of Product Line Extension
First	**Minor change:** Zero to incremental change in store product lines.
Second	**Line extension:** Basic line extension of store product line such as including organic produce, gluten-free products.
Third	**Low-cost souvenir:** Lower-cost addition of product lines such as key chains or postcards.
Fourth	**High-cost souvenir:** Higher-cost addition of product line such as pin ball machines, souvenir artwork, nostalgic products.
Fifth	**Heavy Line-replacement:** Heavy product line replacement carries latest technology.

Figure 10 Source: Author, Ranking
of Product Line Extension.

Because of this discrepancy in product line extensions, businesses which retain their original product format with minor changes in product line extensions have greater iconic value in comparison to businesses with major or heavy product line extensions. There are five Product Line extension categories: Minor change, line extension, low-cost souvenir, high-cost souvenir, and heavy line-replacement. See Figure 10.

Minor-change businesses have minimal changes to their product lines, for example selling one pound and two-pound bags of carrots, or adding new ice cream flavors. Minor-change businesses include Antler's Hotel, Bingley Arms, Angelo Brocato's Confectionery, Brookville Hotel, Bud's Broiler, Cheever's Café, Cinemark Theaters, CVS Pharmacy, Dillard's Department Store, F.M. Light and Sons, Grand Bazaar, Judice Inn, K & B Drugs, La Quinta Inns, Marriott Hotels, McKenzie's Bakery, Mello Joy, Morning Call, Mr. Binky's Adult Shoppe, Nisiyama Onsen Keiunkan Hotel, PJ's Coffee, Pussycat Theaters, Schwegmann's Grocery, Taco Cabana, Walgreen's, WhataBurger, and Whole Foods.

Line-extension businesses split a product line into multiple lines, for example adding ice cream pies and cakes when the original product was ice cream scoops, or older grocery stores increasing their produce lines by adding organic produce into their inventory as opposed to grocery stores which started with organic produce as their inventory.

Examples of Line-extension businesses included Albertson's, A Little Off The Top, Applebee's, Baskin Robbins, Burger King, Dorignac's Grocery, Home Depot, King Sooper's Grocery, Klein's Grocery, McDonald's, Sonic Drive In, Super Cuts, Taco Bell, and Wal-Mart. Low-cost souvenir businesses include $15 or less retail items with state or regional tourist widgets in their inventory. Low-cost souvenir businesses which sell their main product plus disposable mementos such as shot glasses, coffee mugs, or t-shirts and example businesses include: AppleJack Wine & Spirits, Bad Ass Coffee, Café du Monde, Rick's Cabaret, and Starbucks.

High-cost souvenir businesses can include antique stores with place-based and time-based inventory, or higher-priced artisan or general stores. High-cost souvenir businesses have pricier souvenirs such as refurbished old gas station pumps, pinball machines, or rear-ends of old cars converted into love seats. Example businesses include Bellagio, Caesars Palace, Foss Drugs and Irma Hotel, Stardust, and Tropicana. Casinos also sold t-shirts and shot glasses, but inclusion of high-cost souvenirs in their inventory causes casinos to be categorized as high-cost souvenirs.

Heavy Line-replacement businesses are computer, automobile, and stereo businesses. Heavy line-replacement businesses carry the latest technology, have minimal vestige in the past both in their product and store fixtures and are continuously turning over their product lines to carry the newest models of the most recent technology. The conversion of Arap's Old Gun Shop from a gun shop into a bar elicited their heavy product line extension listing. Additional example businesses include: ApplianceWorld, Arap's Old Gun Shop, Best Buy, Chevrolet, Conn's Appliances, and Sprint.

Vice Categories of Businesses

The Human dimension is the last dimension measured in this study and vice is used to quantify the human dimension through measuring the type of vice businesses sell. For this study vices are United States based and centered upon businesses selling age restricted products of alcohol, tobacco, gaming, and nudity/ products. Individual states and counties can regulate these vices, however for this study if one store of a chain carried cigarettes then regardless of store location all stores are viewed as carrying cigarettes. See Figure 11.

It is important to remember humans created vice. In Denver we have news articles about bears plus deer eating fermented berries. For the most part they eat the berries and ultimately sleep off the side effects. Humans created industries based upon providing plus regulating vice.

The five vice categories for businesses are: Outside consumption, In-Store consumption, Hybrid-Innuendo, Multi-Experience, and Sensory-Experience. Outside consumption businesses legally sold vice for outside or non-premise consumption and are mainly grocery, drug, or convenience stores. Example businesses include: Albertson's, Applejack Wine & Spirits, CVS Pharmacy, Dorignac's Grocery, Foss Drugs, K & B Drugs, King Sooper's Grocery, Klein's Grocery, Mr. Binky's Adult Shoppe, Walgreen's, Wal-Mart, and Whole Foods.

Ranking of Vice Sale	Consumption of Vice
First	**Outside Consumption:** Business sells product slated for outside-of-store consumption. Liquor, tobacco stores.
Second	**In-store Consumption:** Business sells product served for in-store consumption. Liquor bars, cigar lounges.
Third	**Hybrid-Innuendo:** Business sells mainstream products served with sexual innuendo.
Fourth	**Multi-Experience:** Business intentionally produces multi-vice combination experience such as liquor and gaming. Sexual innuendo is limited. Casinos (liquor and gaming).
Fifth	**Sensory-Experience:** Business intentionally has topless or nude entertainment as one vice. Gentlemen's Clubs, Strip Clubs, topless or nude Adult entertainment.

Figure 11 Source: Author, Ranking of Vice Dimension.

In-store consumption businesses serve and allow inside-consumption of vice and are usually restaurants, bars, or hotels with liquor licenses to serve liquor within their premises. Example businesses include: Antler's Hotel, Applebee's, Arap's Old Gun Shop, Bingley Arms, Brookville Hotel, Cheever's Café, Irma Hotel, Marriott Hotels, Nisiyama Onsen Keiunkan Hotel, Schwegmann's Grocery (a couple of Schwegmann's stores had lounges near the front of the store), and Taco Cabana.

Outside consumption businesses legally sell vice (liquor/tobacco) for outside or non-premise consumption. In-store consumption businesses serve and allow inside-consumption of vice. Laws governing clothing worn by performers, plus liquor and gaming licenses separate the next three vice rankings. Hybrid-innuendo businesses are traditional non-vice businesses with sexual innuendo and A Little Off The Top is in this category. Multi-experience businesses intentionally serve at least two experiences, such as liquor and gaming, with limited sexual innuendo in comparison to sensory experience businesses. Casinos are a classic example for this category and Bellagio, Caesars Palace, Stardust, plus Tropicana are listed.

Sensory-experience businesses intentionally have topless or nude entertainment as one of their vices. An example of this is prior to the creation of internet and CDs, movie theaters showed reel-to-reel XXX-rated movies. Gentlemen's clubs continue to be open for business and their product line is topless or nude entertainment. Two example businesses in the study are Pussycat Theaters and Rick's Cabaret.

I have no problem converting vice to a point scale. However, there was a problem when adding cumulative points together for each business. When I started to add stigma-points to the cumulative score I observed adding one to five stigma-points to the cumulative globalization index score would not produce a sum number which allowed vice identification. Because of this I used the concept of computer coding to quantify the human dimension. The hundred's column (000, 100, 200, etc.) is for vice values and the last two digits (02, 14, 18, 22, 28, etc) are for the business's summed Globalization index score.

5

Measuring, Explaining, and Diagramming the Six Dimensions

I told you about a peer asking me what part of businesses I was going to measure in quantifying businesses as icons. Now let me explain. During an anthropology course at University of New Orleans our professor told us about how she liked American Indian rugs and she'd bought a video where a rug-weaver gal is being interviewed about her work. The rug-weaver gal is American Indian, so the white guy interviewing her had to go through her son to translate from English to native language and back. And our professor is scribing notes as she watches the film. The white guy asked rug-weaver gal, "How big does the rug become when you are done weaving it?" And the son translates to mom, her response translated via son, "It is about arm to arm length, and whatever height, but it really depends upon life and what is going on, the story I want to tell, what yarn is available, what I am feeling, and the designs I want to weave into the blanket."

Next interview question the white guy asks is, "How hard is it to weave one of these rugs?" The son translates to mom and through him she responds, "It's not that hard because I want to weave the blanket, the weaving has meaning and my life within it, and it is my child's life blanket." The white guy then asks, "How long does it take to weave one of these blankets?" and the response translation is, "It's not really long, but I don't keep track because until the blanket is done, it is not done. So, I don't know. It could take many years, a few years, it really depends."

Our professor looked down at her notes and smirked because the white guy had asked how big, how hard, and how long. Sexual innuendo aside looking at the interviewer's questions we see how the interviewer had not thought through the topic enough to understand how to elicit data and insights from the weaver.

It is important to know the topic you are measuring plus how to measure your topic. My research uses mixed methods of both qualitative and quantitative methods because the number of parking spots a business has or if the business does/ does not have windows or electronic doors does not make the business an icon rather the features which the business carries within itself combine together to make the business an icon.

Hypercube Roles

The existence of the six dimensions were supported in Chapter Two and now it is important to discuss, examine, and list out examples of each dimension. Conceptually, all dimensions are constants divided by change.

How you measure their change is the first kicker. Additionally, these dimensions carry two main roles: The characteristics of what they are plus their position or role as assigned by one's perspective. The characteristics do not change, but roles can change because of differing perspectives. Here is the second kicker. Envision a rotating hypercube and currently Side A of the hypercube is in front of you. Side A has the starring role. Now have the hyper-cube rotate and turn away from facing you so the hypercube is turning toward your right. Because of the rotation your perspective of the hypercube sides have changed and Side B is now in the starring position or role this means Side A has changed roles from being the starring subject and become the secondary object. Here is an example.

When I was looking at the results for stratified proportionate sampling to identify the six dimensions I realized each research paper within the stratified proportionate sampling were snapshots of reality. There were 347 differing snapshots of reality whose insight solidified the existence of entities and constants, plus mutual exclusivity of our six dimensions. Something could be an entity in one research and yet be a constant in another research.

Let's go back to the rotating hypercube for another example. Envision a rotating hypercube to review this chapter's opening story from multiple perspectives. Have one side of the cube in front of you and that is one perspective, then rotate the cube to produce another side of the cube/ another perspective facing you. Specifically, my research of businesses as icons has businesses in the role of entity with theming plus product line extensions in the role of being constants. This rotation allows one dimension to be an entity in one research study and to play the role of a constant or other dimension in another study depending upon the focus of one's perspective.

This English grammar example aids in clarification of mutual exclusivity and roles. In Sentence #1, 'The boy drank the water,' 'The Boy' is the subject and 'Water' is the object in that sentence. Using the six dimensions for Sentence #1 'The boy' is the entity, 'water' is the constant, and 'drank' is the transaction. In Sentence #2, 'The water was consumed by the boy,' 'Water' is the subject and 'Boy' is the object. Boy and Water are entities, but they change places in the subject-role or object-role depending upon one's perspective.

This change in perspective elicits the six-dimensional

hypercube to rotate with Boy facing you as the subject-role (Sentence #1) or Water as the subject-role (Sentence #2) facing you. For my research on businesses as urban cultural icons, businesses are the entity in the subject-role while theming plus product line extensions are the entities in the object-role which makes them constants. Other research papers have theming or product line extensions as the entity because those papers are examining theming or product line extensions. As well with this research being published my studies then become part of the grand data set and becomes a data point of reality.

Measurement Examples to Clarify Each Dimension

Examples of entity include anatomy, animal kingdom, appliances, art, atmosphere, atoms, buildings, businesses, computers, currency, dance, chemical elements, energy, food, music, planets, plant kingdom, stage productions, wavelengths, and viruses. The act of measuring an entity involves quantifying change. For this research I measured my entity businesses by using a hierarchy of human needs (Sarlo 2006a 2006b; 2001; 1992). Using the hierarchy of human needs divides businesses into differing groups. See Figure 12.

Another example of measuring change within an entity is the animal kingdom with the traditional classification hierarchy of phylum, class, order, family, genus, and species as one method to categorize animals into groups. Animals could also be divided into groups by their time of appearance on earth, distance from the North

Pole, whether they are nocturnal versus non-nocturnal, or their food

Measurement Examples	
Entity	Anatomy, animal kingdom, appliances, art, atmosphere, atoms, buildings, businesses, collective conscious, computers, dance, elements, energy, food, housewares, institutions, machines, movies, music, planets, plant kingdom, printed materials, rocks, stage productions, tables, technology, toys, wavelengths, weaponry, and viruses.
Place	Points, lines, paths, volume (i.e., volume in a graduated cylinder, a sphere, a cave), area (L x W x D or X x Y x Z), and polygons.
Time	Age groups, days, dichotomies (i.e., Pre-WWII and Post-WWII), epochs, eras, months, nano-seconds, natural breaks of immigration waves, seasons, and years.
Transactions	Formulas, conversion of energy, human transactions, equations with a relationship between sides such as equal ($=$), unequal ($\neq$, $>$, $<$), or directional. Chemistry examples $Na^+ + Cl^- \rightarrow NaCl$, $PV = nRT$. Physics example $F = MA$. Language examples 'The dog bit the cat,' 'Bob is taller than Walter,' 'Huck hiked into the cave,' and 'Was the battle fought here?'
Constants	Most basic dimension because other five dimensions collapse into being a constant divided various ways.
Human	The human perspective is One's perspective which is defined and refined through interfacing with reality. In contrast social perspectives are also socially defined and created through interfacings with reality. Through cumulative effect both individual plus human groups assign value (0 to +) to everything. Through this value-creation individuals plus groups create and establish their perspectives into customs, culture, beliefs, and norms.

Figure 12 Source: Author, Table of Dimensions
and Examples.

sources. The entity Art could be divided into multiple categories such as Renaissance, Neo-Classic, Romantic, Modern, or Contemporary Art. As well Art could be split into dichotomies such as Romantic v Non-Romantic, Has Red Color v Does Not Have Red Color, or Is Fractal Expressionism v Is Not Fractal Expressionism.

Entities are usually in the Subject-role and are equally at ease in the Object-role.

Examples of place measurement include points, lines, paths, Metropolitan Service Area (MSA) polygons, and GIS measurements of local, focal, zonal, and global (DeMers 2002; Lo and Yeung 2002). Again, how you divide or create change within place is important. Place is where I am right now and where you were five minutes ago. Place is any point along I70 as well as the route of I70 or hiking trails. Places are also cities, MSA's, altitudes, USDA growing zones, lakes, cliffs, or countries. The GIS measurement of place as local, focal, zonal, and global is used in measuring businesses in my research of businesses as urban cultural icons because I combine local, focal, zonal, and global with the concept of cell division to measure the geographical coverage from single-store to multi-national businesses.

Since Place is location then sentences such as 'Go inside the theater,' 'She is inside the car,' 'He went onto the basketball court,' 'The keys are next to the bookcase,' or 'The water filled the drinking glass' have 'inside,' 'onto,' and 'next to' as markers or measurements of place to the extent that the location and area inside of theaters, cars, basketball courts, or drinking glasses are place. The question 'What is the volume inside the flask,' has the word *inside* which indicates place allowing geography to calculate the volume. Since geography measures the area of polygons and volume of oceans, then logic supports the measurement of place's volume inside a graduated cylinder is the same conceptual measurement of place's volume within oceans or seas, caves, or spheres, planets.

Examples of Place in the Object-role include 'Go inside the theater' and 'The keys are next to the bookcase.' Place in the Subject-role occurs when one is comparing different places against each other for example, 'The distance from the Sun is greater for Earth compared to Mars.'

Examples of time measurement include age groups, days of the week and months, dichotomies such as pre-WWII versus WWII combined with post-WWII, or pre-dinosaurs versus dinosaurs plus post-dinosaurs, people under 65 years old versus people 65 years old plus, only people 100+ years old, exact time of births, exact ages, e-pochs, eras, nano-seconds, seasons, or years.

When researchers are surveying people and participants check off their age group then time can be grouped into ages 0 to less than 18 years old, 18 years old to less than 26 years old, 26 years old to less than 36 years old, and so forth. Time can also be repeating dates such as March 14 (3/14 Pi Day), June 28 (6/28 Tau Day), or October 23 (10/23 Mole Day) or singular dates (U.S. Bicentennial July 4, 1976). Depending upon how one divides time, time can be points, ranges, categories, or groups.

To continue with my I70 example time could be when we left Denver, the amount of time to complete our trip, the time we entered Eisenhower Tunnel, or how much time it took to fill my tank of gas.

If our trip had started in Kansas we would also include a Time Zone change and if we drove through parts of Arizona then we might or might not be on Daylight Savings time regardless of the date depending upon whether we are driving through Navajo Nation or not.

When I envisioned pulling out the divider line between pre-dinosaurs versus dinosaurs plus post-dinosaurs and I made time a single grand measurement of one time with no dividers separating time into groups I just stared in dis-belief at the dimensions as I envisioned these six dimensions in their entirety without dividers or change. That was my sudden realization moment because the entirety of the six dimensions without any dividers or grouping is the everything of everything. I stared and realized the six dimensions really are real. Combine this realization moment with realizing the stratified proportionate sampling supported existence of the six dimensions and I kept staring. I went back to the stratified proportionate sampling many times to see my results in order to verify I discovered and tested the six dimensions of everything.

For time in the subject-role compare a full-rotation period around the Sun type-of-year for Mars versus the same measurement for Earth which would produce more Mars-days versus Earth days. This difference in days for a full-rotation year measurement (Time) would be the subject-role and Mars plus Earth (Places) would be in the object-role.

The Constant dimension clarifies that depending upon one's

research perspective that which is a constant in one research and has the object-role can be an entity (subject-role) in another research. This is where envisioning the rotating hypercube comes in handy. Taking a mental picture is important because for that specific moment, research, or idea, that entity is the Belle of the Ball with constant in a secondary position. As well recognize since there are an infinite number of constants it is easier to group them under the constants heading. Returning to the I70 example if your primary focus is terrain flow, then the deer, sheep, trees, native grasses, Colorado River, and I70 would be constants.

Another interesting characteristic of constants is observed when taking apart the Chemistry transaction, PV = nRT. From the constants perspective this transaction uses constants primarily specialized from within Chemistry. Pressure is P, volume is V, n is the number of moles, R is the gas constant, and T is the temperature. Conceptually similar to Theming and Product Line Extensions being under the Constants heading each of these Chemistry constants are separate specialized dimensions also listed under the Constants heading. PV = nRT uses data point values at the ratio level or numbers, versus ranges or dichotomies to calculate the transaction. Constants also clarify Time, Transaction, Place, Entity, and Human are dimensions used in all academic schools of thought and groups of constants are usually limited to particular topics.

To me the Transaction dimension is exciting because this dimension shows the relationships between A and B and/ or the conversion of energy between A and B. Transaction measurement ex-amples include formulas, conversion of energy, human transactions, or relationships/ equations with relationships between sides such as equal (=), unequal ($\neq$, >, and <) or directional arrows ($\rightarrow$ and $\rightleftarrows$).

What makes transactions unique is their ability to relate within or between other dimensions, no other dimension does this task. Since transactions contain at least one dimension on either side of the transaction, then transactions are the connective thread within or between dimensions. Research fields of Transactional, Discourse, Literary Discourse, and Conversational Analysis in sociology, psychology, and anthropology examine one way plus multi-directional transactions.

It is not that we live in a bland six dimension universe. We live in a Five Dimension plus One Über-Dimension universe. Ends up the unique sixth dimension, Transactions, are not a basic dimension, rather Transactions ascends above the five dimensional bailiwick and is the connective thread/ dotted line Über-dimension that trans-pierces and bonds within and between the other five dimensions. Transactions are the capstone and are much more important than the other dimensions because Transactions provide the relationships between dimensions and they are the glue which connects within and between time, place, entity, constants, and human dimensions.

Transactions create and give us our cup of coffee with a little bit of cream and biscotti on the side, our ability to read writing, the conversion of sunlight into plant energy, the ability to step on the gas pedal and have fuel converted into speed or acceleration, humming-birds to fly, water to boil, and sailboats to sail. Think of a movie scene with a bar room brawl, which is a series of transactions, and then mentally list the number and type of transactions you envision. I chuckle when I watch coffee shop scenes in films because trans-actions abound: Waitresses are pouring coffee, the line cook is grill-ing food, people are being seated in booths, other people are ordering their meals, and some people are eating. Then the second hand moves around the clock and a few people get up to pay their tab. I also use these six dimensions as conversational tools when I am talking with people. Pick one of the dimensions they mentioned and use that dimension to turn and expand the conversation.

I had listed transactions as an equal to the other five dimen-sions because I focused on the relationships within transactions and I thought the equal, unequal, greater than, lesser than, and directional arrow signs were the exclusive characteristic within transactions which qualified them as a dimension. The equal, greater than, lesser than, and directional arrow signs are an exclusive characteristic to transactions, but the more exclusive characteristic is that transactions provide the connection between dimensions. Without transactions we have time, place, entity, constants, and human dimensions and no-thing to connect within and between the dimensions.

This ability to be the relationship between and within dimensions nudges transactions into an Über-Dimension, but this distinction is only an identification of uniqueness and is not an 'I am the boss' power elevation. From this I figured out while the Grand five dimension group-label is Dimensions, but Transactions are separate from dimensions as the stand-alone ne plus ultra connection within and between the dimensions. In Figure 13 Transactions are the very dark grey vertical dimension which trans-pierces all other dimensions. See Figure 13.

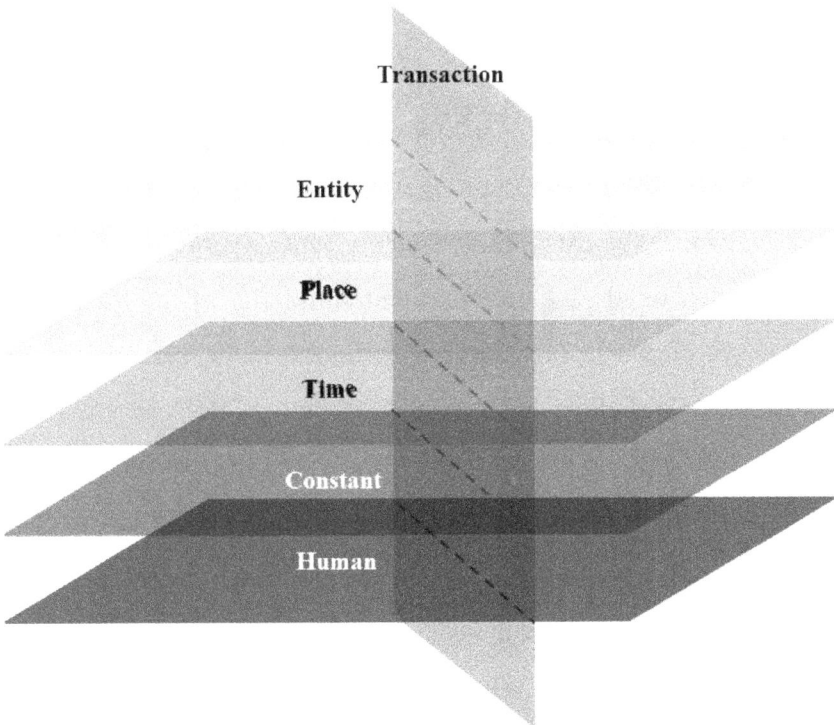

Figure 13 Source: Author Conceptual Framework of
Transactions Intersecting Five Dimensions.

Transactions are similar in concept to using a Horn and Hardart Automat from the 1900s in that transactions ascending above the other five dimensions by being the connective thread/ dotted line columnar dimension which allows the relationships or bonds within and between the other five dimensions. In the literal sense instead of choosing to buy a slice of pie or meatloaf from a specified row and column identified-cell of a Horn and Hardart vending machine, transactions facilitate the relationship between a specific cell within a specified dimension and at minimum another cell in the same or another dimension via the transaction symbol/ terminology. A couple of Chemistry transaction examples are $Na^+ + Cl^- \rightarrow NaCl$ and $PV = nRT$, two physics examples are $F = MA$ and $Speed = D/T$, and language examples are 'The dog *bit* the cat,' or 'Bob *is taller* than Walter,' 'Huck *hiked* into the cave,' and 'Was the battle *fought* here?'

Dimensional plus measurement simplicity of the transaction $Speed = D/T$ allows us to observe the transaction mechanism using D as measurement of place and T as measurement of time. From my I70 example, transactions are the friction between vehicle tires and the highway, the lighting and air ventilation systems within Eisenhower Tunnel, the conversion of fuel into speed or acceleration, plus plant, animal, and human mitochondria converting fuel into energy via Krebs Cycle. Transactions are also mowing lawns, getting married, flying airplanes or gliders, reading the mail, conversations, mud slides, and atomic explosions. As for transactions in the role of entities I am confident research has compared at least a couple of formulas against each other.

Measuring the human dimension was tougher than I imagined. I wanted to say anything human is included in the human dimension, but in Anatomy class we dissected sheep eyes and I can envision someone writing a paper comparing sheep eyes against human, dog, cat, bird, snake, and fish eyes. I nixed human anatomy from the human dimension. Physiology is also nixed because physiology is transactions.

Since the authors of the published journal papers used in the data set are human, I chose the first-person perspective as my example for the human dimension. This is a powerful perspective because the first-person perspective is the direct debriefing from the actor and is created by the individual defining plus refining their knowledge. Their knowledge can include humans assigning value (0 to +) to anything, via customs, culture, norms, and beliefs. Specifically, this example of human measurement of their perspective supports reality being both individually plus socially defined (Berger and Luckman 1966).

This means when I used Mr. Binky's Adult Shoppe as my research spike and stigma came through as a belief quantified into law, my observation allowed me to use stigma to measure the human dimension. Reality being socially defined also means we could have assigned any word for any of the six dimensions, fortunately Penguin,

Orange, and Table were already taken.

The first-person perspective can be enlarged through using a satellite-overview to observe what the first-person perspective includes and does not include. Two other methods of including other perspectives and grouping multiple first-person perspectives expands from having one first-person perspective into having multiple first-person perspectives via surveying, focus groups, or collective groups.

We then use various methods to measure each of these dimensions and once you understand how time can be measured as age groups on a survey, your current age, a dichotomy of all time before dinosaurs versus dinosaurs and later, or your life ten minutes prior when you believed you lived in a three dimensional world, then you can understand there are various methods to measure time, place (i.e., longitude and latitude, polygons, and Einstein's L x W x D) and all the other dimensions.

6

Tehran's Grand Bazaar is the Most Local Cultural Icon

The Guttman measurement scale I formatted is similar in concept to surveying people on a topic, i.e., Love, and the survey has X number of sentences for participants to scale how much they Strongly-Disagree to Strongly-Agree with each sentence and then sum their points. One sentence would discuss caring, another sentence would discuss compassion, and usually there are some negative sentences for participants to mark Strongly-Disagree. Take that concept and apply it to me theoretically surveying businesses on their iconic value using time, place, transaction, entity, and constant.

The Grand Bazaar in Tehran is the true zero for a business as a cultural icon because there are no recorded businesses older than the Grand Bazaar. This means the Grand Bazaar is the oldest recorded operational business. The Grand Bazaar in Tehran started around 1600 BCE and is similar in concept to $0°$ Rankine (R), where any value above $0°$ R involves molecular movement. By default, both the Grand Bazaar and $0°$ R are zero baselines from which measurement occurs.

This means any documented business relative to Grand Bazaar would have an increased globalization score compared to Grand Bazaar. If businesses sold products other than food, or businesses with a more-recent start year meaning they had less history of trading with the community, or if the businesses were more geographically expanded than one continuous location, or used countertop/ desk or conveyor for transactions, or had disjointed theming, or a heavy product line replacement, then those businesses would have an increased globalization index score compared to the Grand Bazaar.

Globalization Index scores merit explanation. The overall concept of globalization index scores is to identify businesses into four categories of Local, Focal, Zonal, and Global Cultural Icons and then identify urban mix ratios. Do not look at the globalization scores and assign a grade to businesses based upon their scores so Starbucks' higher score makes Starbucks less-than an icon compared to Grand

Bazaar. Grading businesses is not the intent of this research. All businesses are icons. Rather the intrigue is to see what image of a culture is being drawn by the icons we choose to trade with through 10-year or 25-year slices of time. This research is a tool to identify the change in culture and to see six dimensions being measured.

Those three outside the U.S. businesses are included because of their longevity since the establishment of documented United States business would begin in 1776. I did not travel to those businesses to triangulate their existence nor physically verify their location, transaction type, etc. The three cases included in this data set are: Bingley Arms in Britain (established in 953, currently open 1060 years), Nisiyama Onsen Keiunkan Hotel in Japan (established in 705, currently open 1308 years), and The Grand Bazaar in Tehran (established around 1660 BCE, currently open 3673 years).

The ending reference year for this study is 2013. If businesses closed before 2013 then the year businesses ended was their final year. For businesses (K & B Drugs, McKenzie's Bakeries, Foss Drug, or Stardust) which closed years before the end-year in this research their operational time was calculated by subtracting year closed from year opened.

This research format is a Guttman scale which is similar to tests professors give to students with tests having cumulated scale-based numeric grades converted into alphabetic-scale grades. It is normal for sectional units of measurement to vary between sections because individual scales are standardized to be combined into cumulative index scores. The Guttman scale concept is the same as five of one-point true or false questions combined with a 5-point fill in the blank and a 15-point essay produce a 25-point grading scale.

This research uses all dimensions with businesses as entity, years in operation as time, geographical distribution of stores as place, type of transaction as transaction, two constants (theming and product line extension), and vice as the human dimension. While each dimensional measurement will have differing amounts of categories the assigned value to the start category for each dimensional measurement begins with 0 (zero). Control groups of businesses were not included because once a business exists it is measurable using all dimensions.

Globalization index scores are summation values for each business using their current dimensional values. Each business is a data point within five dimensions and two constants and their calculated cumulative index scores could range from zero to 534. For this research the data set scores ranged from 000 through 028, so reliability is good because the first digit is the vice value and the last two digits are the primary index score. One case has a value of 000 and to have the calculated maximum value of 534 the business would have to rank as sensory-experience level for vice, a transportation/ communication/ entertainment business, have locations in over 65 countries, between 2,501 and 5,000 years in operation, use a conveyor belt, have a disjointed theme and heavy product line replacement.

Adding vice's stigma-points into the cumulative globalization index score would not yield a sum value allowing identification of business vice, so I applied the concept of coding where the placement of numbers identified and quantified the human dimension. All businesses have a three digit score and the hundred's column (000, 100, 200, etc.) is vice values with the last two digits (02, 14, 18, 22, 28, etc.) are business's summed Globalization index score. Using the Irma Hotel's index score of 209 indicates inside consumption vice with a local cultural icon because of the two right column values of 09 as part of the score.

Adding Up Businesses Numbers

Let's first discuss scores and the businesses, then we will examine business types and their scores. The Grand Bazaar in Tehran has the lowest index score (000 points) because venues within the Bazaar sell food, the bazaar remains at its original location, is extremely old, it has verbal transactions where the employee used more energy compared to the customers, minimal theming, minor product line extension, and no vice. Nisiyama Onsen Keiunkan Hotel has the next lowest index score (203 points) because the hotel sold shelter, was not as old as the Grand Bazaar, had blended theming, and sells liquor for inside consumption.

Other low index score businesses are: Judice Inn (burgers),

F.M. Light and Sons (clothing), Brocato's Confectionery, and Morning Call Coffee with 005 index score. Bingley Arms has the same index value of 005 but sells liquor for inside consumption, so their index score is 205. See Figure 14 for local cultural icon businesses with globalization scores ranging from 0 through 209.

Antlers and Brookville Hotel are older businesses with slightly higher entity values, plus higher theming value so these two businesses score 006 and 007 with 200 vice level. Other low index scoring businesses include a coffee shop, burger joint, grocery stores, and Buffalo Bill's Irma Hotel in Wyoming all having the same business score of 009. Dorignac's Grocery and Klein's Grocery plus Irma hotel have 100, 100, and 200 vice scores.

The interplay between place, time, and transactions create a tight five-point cluster with ten cases. Having a slightly larger geographical expansion, being newer businesses, and expanded transacttions has Schwegmann's Grocery (Mesotype $Focal_2$ grocery store), McKenzie's Bakery (Intertype $Focal_2$ bakery), Foss Drugs (Intertype $Local_0$ outside consumption drug store), Applejack Wine & Spirits (Intertype $Local_1$ conveyor belt transaction outside consumption liquor store), Arap's Old Gun Shop (Intertype Heavy replacement product extension bar), A Little Off The Top (Protype hybrid experience hair salon), PJ's Coffee (Archetype $Zonal_1$ coffee shop), Mr. Binky's Adult Shoppe (Protype $Focal_1$ outside consumption adult shoppe), Taco Cabana (Archetype $Zonal_1$ fast food), and Tropicana (Archetype

	Business type	Place	Time	Trans-ation	Theming	Product Extension	Human	Index Score
								LOCAL CULTURAL ICON
Grand Bazaar	Food	$Local_0$	0	Verbal	Minimal	Minor	000	000
Nisiyama Onsen Keiunkan Hotel	Shelter	$Local_0$	1	Verbal	Blended	Minor	200	203
Judice Inn	Food	$Local_0$	5	Verbal	Minimal	Minor	000	005
F.M. Light and Sons	Clothing	$Local_0$	4	Verbal	Minimal	Minor	000	005
Brocato's	Food	$Local_1$	4	Verbal	Minimal	Minor	000	005
Morning Call	Food	$Local_1$	4	Verbal	Minimal	Minor	000	005
Bingley Arms	Entertainment	$Local_0$	1	Verbal	Blended	Minor	200	205
Antlers Hotel	Shelter	$Local_0$	4	Verbal	Blended	Minor	200	206
Brookville Hotel	Food	$Local_1$	5	Verbal	Blended	Minor	200	207
Cheever's Cafe	Food	$Local_0$	7	Verbal	Blended	Minor	200	208
Mello Joy	Food	$Focal_0$	5	Verbal	Blended	Minor	000	009
Bud's Broiler	Food	$Focal_1$	5	Verbal	Blended	Minor	000	009
Dorignac's Grocery	Food	$Local_1$	5	Conveyor belt	Minimal	Line extensions	100	109
Klein's Grocery	Food	$Local_1$	5	Conveyor belt	Minimal	Line extensions	100	109
Irma Hotel	Food	$Local_0$	4	Verbal	Fused	High cost	200	209

Figure 14 Source: Author, Local Cultural Icon
Globalization Index Scores.

$Local_0$ entertainment) grouped near each other with index scores of 110, 011, 112, 113, 213, 313, 014, 114, 214, and 414 respectively. The difference between PJ's Coffee, Mr. Binky's Adult Shoppe, Taco Cabana, and Tropicana are their vice levels.

Five businesses with index scores ending in --15 all have vice with the lowest scores being a grocery and drug store (closed), followed by two casinos (2nd casino closed) and an XXX-rated adult film theater (closed). These businesses are King Soopers (index score of 115), K & B Drugs (index score of 115), Caesar's Palace (index score of 415), Stardust (index score of 415), and Pussycat Theaters (index score of 515). See Figure 15 for data set second quarter displaying focal cultural icon businesses with globalization scores ranging from 110 through 515.

Opening up in 1905 did little to aid K & B Drugs with an index score of 115. Rather, the difference of Foss Drugs being $Local_0$ and K & B Drugs being $Zonal_0$ explains the increase in three points. WhataBurger started 1950, expanded into Mexico ($Global_0$), and is only one point higher (index score of 016) than K & B Drugs (See Figures 15 and 16). Albertsons (index score of 116) has the same base score as WhataBurger (index score of 116) and Bellagio (index score of 416). Café du Monde opened over one hundred fifty years ago, but their global expansion and low-cost souvenirs results with an index score of 017. Café du Monde and ApplianceWorld (index score of

	Business type	Place	Time	Trans-action	Theming	Product Extension	Human	Index Score
FOCAL CULTURAL ICON								
Schwegmann's Grocery	Food	$Focal_2$	4	Conveyor belt	Minimal	Minor	100	110
McKenzie's Bakery	Food	$Focal_2$	5	Counter top	Minimal	Minor	000	011
Foss Drugs	Health care	$Local_0$	5	Counter top	Minimal	High cost	100	112
Applejack Wine & Spirits	Entertainment	$Local_1$	5	Conveyor belt	Minimal	Low cost	100	113
Arap's Old Gun Shop	Entertainment	$Local_0$	5	Verbal	Blended	Heavy re-placement	200	213
A Little Off The Top	Personal care	$Focal_0$	7	Verbal	Blended	Line extensions	300	303
PJ's Coffee	Food	$Zonal_1$	6	Verbal	Blended	Minor	000	014
Mr. Binky's Adult Shoppe	Entertainment	$Focal_1$	7	Counter top	Minimal	Minor	100	114
Taco Cabana	Food	$Zonal_1$	6	Verbal	Blended	Minor	200	214
Tropicana	Entertainment	$Local_0$	6	Verbal	Fused	High cost	400	414
King Sooper's Grocery	Food	$Zonal_0$	6	Conveyor belt	Minimal	Line extensions	100	115
K & B Drugs	Health care	$Zonal_1$	5	Counter top	Minimal	Minor	100	115
Caesars Palace	Entertainment	$Local_0$	7	Verbal	Fused	High cost	400	415
Stardust	Entertainment	$Local_0$	7	Verbal	Fused	High cost	400	415
Pussycat Theaters	Entertainment	$Focal_2$	6	Verbal	Fused	Minor	500	515

Figure 15 Source: Author, Focal Cultural Icon Globalization Index Scores.

117

017) have the same index score as well the same base score of CVS Pharmacy (index score of 117). See Figure 16 for zonal cultural icon businesses with globalization scores ranging from 016 through 520.

Dillard's Department Store, Sonic Drive In, Walgreen's, Taco Bell, and Baskin Robbins are close enough in their place, time, transactions, and theming values for their base score to be −18 or −19. Within this group Dillard's Department Store, Sonic Drive In, and Walgreen's had the lowest value of 018, with Walgreens having 100 vice value (index value of 118). Both Taco Bell and Baskin Robbins have the same scoring values (index scores of 019) across the scale because they are Food Global$_1$ Intertype verbal transactions, blended theming, and line extensions.

Opening their business in 1895 could not aid Conn's Appliances in retaining a low cultural index score. Conn's Appliances joins La Quinta Inns and Suites and Rick's Cabaret with a base score of −20 points because being an appliance store selling cutting edge technology increases Conn's Appliances score.

For vice Pussycat Theaters (515 index score) and Rick's Cabaret (520 index score) have the highest vice scores (5--) with Rick's Cabaret slightly more geographically expanded compared to Pussycat Theaters.

	ZONAL CULTURAL ICON							
	Business type	Place	Time	Trans-action	Theming	Product Extension	Human	Index Score
WhataBurger	Food	$Global_0$	5	Verbal	Minimal	Minor	000	016
Albertson's	Food	$Zonal_2$	5	Conveyor belt	Minimal	Line extensions	100	116
Bellagio	Entertainment	$Local_0$	8	Verbal	Fused	High cost	400	416
ApplianceWorld	Appliances	$Focal_1$	7	Verbal	Blended	Heavy re-placement	000	017
Café du Monde	Food	$Global_0$	4	Verbal	Minimal	Low cost	000	017
CVS Pharmacy	Health care	$Zonal_2$	6	Counter top	Minimal	Minor	100	117
Dillard's Department Store	Clothing	$Zonal_3$	5	Counter top	Fused	Minor	000	018
Sonic Drive In	Food	$Global_0$	5	Verbal	Blended	Line extensions	000	018
Walgreen's	Health care	$Zonal_4$	4	Counter top	Blended	Minor	100	118
Taco Bell	Food	$Global_1$	5	Verbal	Blended	Line extensions	000	019
Baskin Robbins	Food	$Global_1$	5	Verbal	Blended	Line extensions	000	019
Conn's Appliances	Appliances	$Zonal_0$	4	Counter top	Disjointed	Heavy re-placement	000	020
La Quinta Inns	Shelter	$Global_0$	6	Verbal	Fused	Minor	000	020
Rick's Cabaret	Entertainment	$Zonal_1$	6	Verbal	Fused	Low cost	500	520

Figure 16 Source: Author, Zonal Cultural Icon
Globalization Index Scores.

119

The first few Global Cultural icon businesses are Super Cuts, Home Depot, Whole Foods, Marriott Hotels, and Applebee's restaurants with base scores of –21. See Figure 17 for global cultural icon globalization index scores ranging from 021 through 028.

McDonald's, Burger King, Wal-Mart, Bad Ass Coffee, Cinemark Theaters, Starbucks, Sprint Communications, Best Buy, and Chevrolet are higher scoring businesses with index scores of 022, 022, 122, 023, 023, 024, 025, 025, and 028 respectively. The calculated scale maximum is 534 and the closest case within this study is Chevrolet (started in 1911, in over 140 countries, with heavy product line replacement). The index score for Best Buy (025) is near Chevrolet's index score but in contrast to Chevrolet being transportation and very global, Best Buy is appliance-based and younger.

Local, Focal, Zonal, and Global Cultural Icons

These Index scores are divided into four groups and are premised upon weighing the index scores against their geographical scores. The categories are: Local Cultural Icon with scores ranging from 000 to 509, Focal Cultural Icons with scores ranging from 010 to 515, Zonal Cultural Icons with scores ranging from 016 to 520, and Global Cultural Icons with scores ranging from 021 to 534.

	Business type	Place	Time	Transaction	Theming	Product Extension	Human	Index Score
GLOBAL CULTURAL ICON								
Super Cuts	Personal Care	$Global_0$	6	Verbal	Blended	Line extensions	000	021
Home Depot	Essential furnishing	$Global_0$	6	Counter top	Minimal	Line extensions	000	021
Whole Foods	Food	$Global_0$	6	Conveyor belt	Fused	Minor	100	121
Marriott Hotels	Shelter	$Global_2$	5	Verbal	Fused	Minor	200	221
Applebee's	Food	$Global_1$	6	Verbal	Fused	Line extensions	200	221
McDonald's	Food	$Global_4$	5	Verbal	Blended	Line extensions	000	022
Burger King	Food	$Global_4$	5	Verbal	Blended	Line extensions	000	022
Wal-Mart	Personal Care	$Global_1$	5	Conveyor belt	Minimal	Line extensions	100	122
Bad Ass Coffee	Food	$Global_0$	7	Verbal	Disjointed	Low cost	000	023
Cinemark Theaters	Entertainment	$Global_1$	6	Verbal	Fused	Minor	000	023
Starbucks	Food	$Global_3$	6	Verbal	Fused	Low cost	000	024
Sprint	Communications	$Zonal_4$	6	Verbal	Fused	Heavy replacement	000	025
Best Buy	Appliances	$Global_0$	6	Counter top	Blended	Heavy replacement	000	025
Chevrolet	Transportation	$Global_4$	4	Verbal	Fused	Heavy replacement	000	028

Figure 17 Source: Author, Global Cultural Icon
Globalization Index Scores.

121

All grocery stores sold vice with three grocery stores as local cultural icons: Dorignac's Grocery (109 index value) currently open for business, and the other two (Klein's Grocery with 109 index value and Schwegmann's Grocery with 209 index value) are permanently closed. King Soopers (115 index value) is a focal cultural icon and Whole Foods (121 index value) is a global cultural icon.

Because two local cultural icon grocery stores are permanently closed, there is a poor prognosis for grocery stores as lo-cal cultural icons to remain open. A lack of family interest in contin-uing local cultural icon grocery stores (usually family-based) is a po-tential factor. In contrast, focal through zonal cultural icon grocery stores usually are not family-run businesses so they can have larger fiscal budgets to remain open for business. From this I can envision focal, zonal, and global cultural icon grocery stores being the bigger dinosaurs and the local cultural icon grocery stores look like they are going to be eaten alive by the big boys.

All four drug stores sold vice with two of the drug stores as focal cultural icons and both are permanently closed (Foss Drugs with an index value of 112 and K & B Drugs with an index value of 115). In contrast other two drug stores are zonal cultural icons and also remain open for business (CVS with an index value of 117 and Walgreens with an index value of 118). Walgreens has an inter-national counterpart of Alliance Boots but their names have not blend-ed to create an international Walgreens chain potentially because of international prescription regulations plus cultural differences in products.

When it came to burgers the four cases are distributed across three cultural icon levels. Judice Inn (index value of 003) and Bud's Broiler (index value of 009) are local cultural icons, WhataBurger (index value of 016) is a zonal cultural icon, and McDonald's (index value of 022) is a global cultural icon, so burgers are crossing national borders and becoming global cultural icons. Of the two Mexican fast food businesses in the study, Taco Cabana has an index score of 214 (focal cultural icon) and Taco Bell has an index score of 019 as a zonal cultural icon which is interesting because Mexican food has crossed national borders but Mexican fast food businesses have not crossed national borders.

Coffee shops range from local to global cultural icons with Morning Call (index score of 005) and Mello Joy (index score of 009) as local cultural icons, PJ's Coffee (index score of 014) is a focal cultural icon, Café du Monde (index score of 017) is a zonal cultural icon, and both Bad Ass Coffee (index score of 023) plus Starbucks (index score of 024) are global cultural icons. The power of independent coffee shops is strong, so Bad Ass Coffee and Starbucks ranking as global cultural icons is interesting.

There is now a Starbucks at Vatican City and it will be interesting to observe the customer base demographics. Since customer purchases keep stores open, I am intrigued if more United States citizens are wanting a cup of familiar coffee their taste buds will recognize and/ or are Italians and others wanting a United States-based Starbucks brand experience. Bad Ass Coffee had overseas locations when I studied them but since that time their locations are now United States based so if the end date for this study was 2017 their index score might have been different.

Within the Entertainment category, eleven businesses range from local through global cultural icons. Bingley Arms in Britain is the only local cultural icon (index score of 205), seven businesses are focal cultural icons (Applejack Wine & Spirits with index score of 113, Arap's Old Gun Shop with index score of 213, Mr. Binky's Adult Shoppe with index score of 114, Tropicana with index score of 414, Caesar's Palace with index score of 415, Stardust with index score of 415, and Pussycat Theaters with index score of 415). Bellagio (index score of 416) and Rick's Cabaret (index score of 520) are zonal cultural icons, and Cinemark Theaters (index score of 023) is a global cultural icon.

From these findings it is observed that being known globally does not necessarily equal being a global cultural icon, as well, having a Global place ranking does not necessarily mean one will be a global cultural icon. Since Colorado is a legal marijuana state I could have included retail marijuana stores under the entertainment heading, but these businesses would have not added new insights because their vice level would be Level One Outside Consumption and they would be limited in geographic expansion to the borders of Colorado. More states have legalized marijuana which means a future study could include these businesses.

Over half the businesses in this study have a Human dimension level of 1 or higher with the majority selling 'outside-the-store' consumption. More interesting is Wal-Mart having the greatest globalization score with a high index value plus a vice value at 122. High value vice businesses such as Hybrid Innuendo, Multi Experience, and Sensory-Experience did not have place values above $Zonal_1$.

The Playboy Club was multinational and all three waves of clubs (first wave was 1960 through 1991, second wave was 2006 through 2014, and third wave of clubs was 2014 through current) had gaming tables, liquor, and bunny uniforms, but they did not cross my mind to include them into the data set when selecting businesses for the data set. I would have listed Playboy Club by their three waves (Playboy Club I, Playboy Club II, Playboy Club III) and their vice would have been Multi-Experience. Their respective index scores would have been 324, 325, and 325. The Playboy Club brings up a valid point to review the strengths plus weaknesses of this research, future opportunities, applications of their research, make conclusions and suggestions.

To make this research stronger more protype business cases, especially global-protype need to be included in the data set. As well, more supratype, ultratype, and pantype businesses (501 to 1000 years old, 1001 to 2500 years old, and 2501 to 5000 years old respectively) need to be included in the data set for comprehensive case comparison. Hamleys is the world's oldest operational toy store with 167 stores in 18 countries and would be $Global_2$ Entertainment Mesotype business. Hamleys began as 'Noah's Ark' in 1760 (BBC 2010), but I cannot identify when in the business's longevity the business changed names and became operational as Hamleys. Another Category Four transportation entity business to include is trains/ railroads. Four popular $Global_0$ trains are: Amtrak covering two countries, Eurostar plus Thayls cover four countries, and London Venice Simplon Orient Express trains pass through five countries. Trains would have to cover nine or more countries to be $Global_1$.

A recent datum of a crabbing boat converted into a floating strip club bar in Alaska, operating during 2014 (Andrews 2014), would be interesting because of its Local measurement. Another discussion of local meriting further examination is Constance, Harvel, and Prelog's (2017) presentation of "Tennessee Whiskey as Terroir?" at SSSA 2017. Sugar and corn used in Tennessee Whiskey fermentation process are not necessarily grown in the state where Tennessee Whiskey is distilled, so is Tennessee Whiskey local at the state level because that is where conversion of ingredients to finished product occurs or does ingredient origin weigh into the decision. As well if whiskey is distilled just across the state line in Kentucky, can that whiskey be labeled Tennessee Whiskey if distillers followed Tennessee Whiskey distillation process. For my research Global vice businesses or supra-, ultra-, and pan-age vice businesses would be amazing cases to include in the study.

When it comes to cases I should have included into the data set Service Merchandise and the Disney complex would have been good cases to include. Service Merchandise had conveyor belts and the Disney complex would be Intertype Global$_0$ with a score of 025 and be in company with Sprint and Best Buy.

Categorizing businesses into a range of four urban cultural types is to provide structure in viewing percentage mixes of the four groups within cities or MSA. Time-based change of urban mixes will be observed by mapping index values in 10 or 25-year increments. Initially the ratio of all four cultural icon group in any beginning city should be 100: 0: 0: 0 with all businesses having local iconic values. Cities with equal amounts of the four groups would have 25: 25: 25: 25 ratio and a city with only global cultural icon businesses would have 0: 0: 0: 100 ratio.

I recently visited St. John's Newfoundland and in 1892 St. John's had a fire in the commercial area near their water-front (Heritage 2007). The fire burned down a couple of streets ma-king 11,000 people homeless and causing $13 million in damages. Yellowbelly's brewery & public house was one of many gutted businesses and Yellowbelly's was rebuilt and remains open plus oper-ational since 1725. The impact of this fire on St. John's ratio of businesses as urban cultural icons would be interesting to map. Two prior fires in St. John's occurred during the 1800's, so these fires deforested/ wiped clean the urban landscape changing St. John's cultural icon group ratios from the day prior to the fire.

Through time ratio values will evolve and between-city comparisons of cities with differing index score ratios can then be compared to observe for strengths, opportunities, opportunities, and weaknesses of differing index score ratios. Older more established city should have a higher percentage of local cultural icon businesses; in contrast MSA's with a relatively recent influx of growth might display more global cultural icon businesses. Focal cultural icons tend to occur when local cultural icon businesses expand to two or more locations within MSA's and zonal cultural icon businesses occur when state lines are crossed for expansion. The comparison of cultural iconic ratio-percentages of cities will provide insights on cultural change, business geography, retail businesses, and globalization patterns within cities.

Urban planners can use this matrix to guide business mixes between the four cultural iconic groups to have more businesses with local cultural icon scores or towards having more businesses with other cultural icon scores. The economics of traditional brick and mortar businesses are changing and this matrix allows cities to be flexible in their response.

Weaknesses in This Research

This study establishes and uses six dimensions to quantify businesses as totems which can measure the changing culture of a city via index scores. The data set needs to expand to 350+ cases for robust statistical testing and to allow entire cities to be mapped. Lodging next. As well, drive-in theaters need to be included in the data set to contrast against Pussycat and Cinemark Theaters. More telecommunication, transportation (train and bus stations, airport and entertainment businesses need to be included in the data set. Studebaker and other types of transportation, maybe trains. Service businesses such as medical, tax preparers, accountants, lawyers, and health care need to be included in the data set. Changes in the purchasing of office supplies from brick and mortar businesses to internet businesses plus the subsequent consolidation and buy outs of office supply businesses need to be observed.

I thought of historically Black Colleges and Universities and that made me think of historically Black businesses. A comparison at various time increments of local through global cultural icon ratios of historically Black businesses within West Coast, Southern, Northeast, and Midwest cities should elicit interesting patterns. As well, ratio changes of Hispanic businesses in border and non-border cities plus ratio changes of women-owned businesses in varying cities should elicit in economic growth insights of these groups. I did not think to including the Playboy Clubs nor include Mustang Ranch in Nevada.

Additional insights on businesses long gone can be pursued via personal histories, historical centers, telephone books and Dunn's Registry.

Changes Since This Research

While this research project ended in 2013 some changes on the business playing field have occurred. CVS pharmacy is now inside Target stores which might change CVS's place-ranking, plus Starbucks anticipates doubling their store count with an additional 12,000 stores built by 2021. Godiva plans on opening 2,000 cafes by 2025, and SeaWorld is expanding to Dubai in 2020. Safeway groceries bought Albertson's in 2014 and a few Albertson's stores remain because of anti-trust requirements so Albertson's index score might decrease because of less locations.

When it comes to e-commerce Sears (Jones 2017) plus Macy's (Macys 2016) announced poor fourth-quarter performance of their brick-and-mortar stores necessitating store closures, while their e-commerce is strong. E-commerce transaction taxes are based upon consumer's zip code, so $Global_4$ would be their geographical listing. To include e-commerce, transactions would be expanded to a fourth ranking of 'Employee receives the order via internet' and e-commerce theming could be measured by the structure, logic, and formatting of websites. Line extensions for e-commerce would rank niche businesses with a lower value and Amazon or Zappos higher.

Discussing these individual cases plus the four geographically based icon groups brings up my peer's question asking how I planned on measuring businesses. And that brings a family story to mind. Back in the early 80s we received the phone call my grandmother had about a week to live and we drove to Indianapolis IN to meet my dad, be there for when she died, and to then stay, plan, plus oversee her funeral service. So we've been in Indianapolis for a few days and grandmother was very frail, we were on the short end of her time line. The phone rings around 2:20 AM with the hospital staff saying get to the hospital because we have maybe an hour with her. Dad, my husband, I, and grandmother's two surviving sisters load into the car and head over to the hospital. We've said our good-byes in grand-mother's hospital room and we left her room to sit in the nearby visitor area while the staff did medical stuff with grandmother. About a mi-nute later the newly-minted orderly MD guy walked over to tell us, "I am sorry to inform you but she just died." But before he came over to tell us the news Aunt Birdie (grandmother's sister) took that moment to visit the ladies room, so she missed his announcement. Aunt Grace (the other sister) immediately broke down crying and right at that moment Aunt Birdie opened the door from the ladies room to join up with us and Aunt Birdie was focused on juggling her walker to get the walker legs over the restroom door threshold so she could re-join the group.

Aunt Birdie had perfect hearing, but she just didn't hear what the orderly guy said because again, she was wrestling her walker and the restroom door so she asked, "What?" This must have been the orderly's first death and he failed to read the situation clearly. He thought Aunt Birdie had a hearing problem and she was hard of hearing, so the orderly yells directly into Aunt Birdie's 87 year-old face, "SHE JUST DIED."

We bit our lips to not laugh about his bad timing, but his mistake was to make a false conclusion. I think about that and I connect false conclusions back to my peer asking if I was going to count windows, the number of parking slots were in the business's parking lot, or if the business had electric doors to measure business's iconic values. If I had followed my peer's line of reasoning and counted windows or whatever then this scale and the six dimensions would never have occurred because of my peer's false conclusion. I so severely appreciate George Ritzer, his *Globalization of Nothing* (2004) continuum, and identifying the six dimensions.

Confirmation We are Measuring the Right Stuff

There are add-on statistical tests that can be run as additional validity testing for the six dimensions. Let me lead into some humor to aid in explaining the concept of these statistical tests. This neutron walks into a bar and orders a drink and the bartender says, "For you, no charge" (Unknown 2017), but my husband took a while to memorize the joke so he would mess-up the punchline by saying, "It's happy hour," or "For you, it's double," or "For you, I need to see some ID." And that is the basic premise for factor analysis. The data is there but the punchline isn't necessarily clear. Factor analysis keeps the data in its original position and instead of changing the position of the data, factor analysis rotates the x- and y- axis to find the best fit for the data because maybe the axis are in the wrong position and should not be the traditional horizontal and vertical axes.

Validity testing is used to observe how much overlap or correlation is between variables and validity testing correlation values range from 0 (zero) which is no co-relating between the variables to + 1.00 or -1.00. The closer correlation values are to either +1.00 or -1.00 means a stronger co-relating relationship between two variables. Most researchers want high positive or negative correlation values because their theory is either 'As A increases, then B increases or as A increases, then B decreases' in contrast we want to observe low correlation to support these dimensions being mutually exclusive. Time should not equal Place.

For this research we want to observe low positive or negative correlation values because there should be a minimal correlation between the six dimensions. As well if correlation values are oddly high then my data set of businesses would merit revision because this would indicate my choice of cases for the data set was heavier for particular type of businesses.

Twenty-one validity testing correlation values are listed in

Figure 18 with all values below 0.5 which indicates all relationships are weak and supports Time does not equal Place. See Figure 18.

	Business Type	Time	Place	Transaction	Theming	Product Line
Time	.218**					
Place	−.131	.167				
Transaction	−.022	.022	.031			
Theming	.271**	.291**	.238**	−.397***		
Product Line	.463***	.261**	.008	.012	.390***	
Human	.468***	.249**	−.416***	−.132	.317***	.15

* p<.10, ** <.05, ***p<.01

Figure 18 Source: Author, Correlation Matrix.

There are 21 correlation values listed in Figure 18 with all values below 0.4 meaning all relationships are weak. For this research we want to see weak relationships because Entity does not equal Human dimension.

In this data set the weakest relationship is between Place and Business type at -0.004, so a very faint negative correlation is observed. When the geographical expansion of business types increases in coding value from $Local_0$ to $Global_4$, the categories of businesses decreases in coding from the fourth business category (transportation, communication, and entertainment) down to the first business category (food) which means the larger the geographical expansion, then the more likely the business is food-based. Having correlation value close to 0.000 means minimal relationship between the two

134

dimensions. The data set ranges from:

- Twenty-eight cases ranging from single site $Local_0$ first business category of food (Grand Bazaar in Tehran) up to $Global_3$ first business category (Starbucks).

- Six cases ranging from single site $Local_0$ second business category of shelter and clothing (Nisiyama Onsen Keiunkan Hotel and F.M. Light and Sons) up to $Global_2$ second business category (Marriott Hotels).

- Eleven cases ranging from single site $Local_0$ third business category of health care, personal care, and essential furnishings (Foss Drugs) up to $Global_1$ third business category (Wal-Mart).

- And thirteen cases ranging from single site $Local_0$ fourth business category of transportation, communication, and entertainment (Bingley Arms) up to $Global_4$ fourth business category (Chevrolet).

The strongest relationship is between Theming and Place at 0.374 and this relationship is statistically significant with a probability less than 0.01 that these relationships would randomly occur in reality. The significance of low probability value means this is a statistically significant relationship at 0.374. The 0.374 correlation value between increasing theming categories and increasing geographical expansion is considered a weak relationship. The data set ranges from:

- Twenty cases range from single-site $Local_0$ first theming category of minimum theming (Grand Bazaar in Tehran) up to $Global_1$ first theming category (Wal-Mart). There is a decrease in minimum theming cases as place ranking increases.

- Twenty cases range from single site $Local_0$ second theming category of blended (Nisiyama Onsen Keiunkan Hotel) up to $Global_0$ second theming category (Best Buy). Blended theming is relatively evenly scattered throughout the data set as place ranking increases.

- Sixteen cases range from single site $Local_0$ third theming category of fused (Irma Hotel) up to $Global_4$ third theming category (Chevrolet). There is an increase of fused theming businesses as place ranking increases.

- Ending with two disjointed cases the first case is multi-state $Zonal_0$ fourth theming category of disjointed (Conn's Appliances) and the next case is $Global_0$ fourth zoning category (Bad Ass Coffee).

Overall Figure 18 displays low correlation values using five dimensions plus two constants. Three additional statistics are Determinant = .200, KMO = .487, and Bartlett's Sphericity = 86.63, $p < .001$. The Determinant value is low which supports dimensions are measuring different business characteristics.

KMO (Kaiser-Meyer-Olkin) is a measurement of data suitability for factor analysis and values range from 0.000 to 1.0. The closer the value is to 1.0 then the better match-up between data set and running factor analysis. KMO value of 0.487 is slightly below the cut off value of 0.500, but at the same time the sample size to variables ratio is 58 cases to seven variables (~7.5 cases to one variable) which tends toward supporting running factor analysis with mid-range KMO values.

Bartlett's Sphericity tests for observation that the correlation matrix is an identity matrix with the minimum accepted significance value set at $p < .005$. Because the significance of Bartlett's Sphericity ($p < .001$) for this data set is significantly below the minimum accepted value of $p < .005$ this means Bartlett's Sphericity supports performing factor analysis for structure detection. Three additional criteria support using factor analysis:

- All communalities after extraction are .717 or higher with seven original variables. Communalities are the proportion of the variance of each variable that can be explained by the factors.

- Seven dimensions (two constants) are tested.

- And components retained account for at least 81.9 percent total variability. Overall, these findings support within this data set Time does not equal Place.

		Loading
Component 1	**Open Systems**	
	Product Line Extension	0.897
	Business Type	0.750
Component 2	**Local Selling of Vice**	
	Place	0.861
	Human	-0.776
Component 3	**Facilitating Human Contact**	
	Transactions	-0.911
	Theming	0.705
Component 4	**Younger Businesses sell more Vice plus more Themed Experiences**	
	Time	0.928
	Human	0.379
	Theming	0.345

Figure 19 Source: Author, Components.

The rotated solution of Factor Analysis support six dimensions by revealing four subcomponents of the six dimensions: Open systems, local selling of vice, facilitating human contact, and new businesses sell more vice plus have more theming. Four components account for 79.67 percent of total variability. Component #1 is Open systems and accounts for 22.86 percent of the score variance plus open systems explain the process of businesses introducing change within their structure and is discussed in Chapter Two. See Figure 19.

The local/ focal selling of vice versus global businesses selling vice account for 21.41 percent variance. This relationship between geographically Local Multi-Sensory (Level 4 Vice) and Sensory Experience (Level 5) Vice businesses with geographically Global No-Vice (Level 0 Vice) and Outside Consumption (Level 1) Vice businesses could be based upon stricter laws involving the geographic expansion of vice into other countries, differing indigenous vice levels, and international monies. Also, while I did not include Playboy Clubs or Mustang Ranch into the data set their inclusion into the data set might have decreased the loading value of Place into Component Two.

Facilitating human contact illustrates the relationship between business transactions and theming and account for 19.90 percent variance. The more businesses have verbal transactions, then the more likely business theming increases. In turn this relationship can provide more of a consumer experience. New Businesses sell more vice plus have more theming and account for 17.73 percent of the variance (79.67 percent). On a side note two dimensions loaded on two factors.

7

What Aristotle and Buddies
Bring to the Table

We live in a Six Dimension world and these six dimensions are as contemporary as today plus are an archeological vestige from the beginning. I identified these dimensions in my dissertation and when I discussed them with one of my committee chairs, he asked why these were dimensions when Aristotle's cosmology had identified Earth, fire, water, ether, and air, and Albert Einstein listed his three dimensions. This leads into discussing prior perspectives on dimensions and a tool I use is to stand within other perspectives to combine gestures with written and verbal words to envision their perspectives. I also assess the emotions I am feeling because most likely those emotions are what the other perspectives felt.

Strengths and Weaknesses of Other Perspectives

Ancient Greeks measured reality using a different perspective compared to my dimensions. While earth is entity or place depending upon Aristotle's definition, water, ether, air, and fire could be entities with water and fire as transaction (i.e., water plus fire = steam), plus fire needs air to exist so linear sequence exists. However, Aristotle did not directly acknowledge time, place, human or constants, plus he did not specifically use 'transactions.'

Another prior perspective merits reexamination. The three dimensions of X x Y x Z axis merit reexamination because X x Y x Z as three dimensions does not allow for all geographic, time, transactional, entity, constants, or human dimensional measurements to exist. Specifically, Einstein did not labeled Time as a dimension, plus he did not label X x Y x Z as one of many types of geographical measurement (i.e., points, lines, polygons, and GIS local/ focal/ zonal/ and global).

Because of this discrepancy Albert Einstein's (1961/ 1952, 1920, 1916, 1905) three dimensional reality is limited and incomplete. Figure 4 in Chapter Two supports the six dimensions exist and fails to support Einsteinian three dimensions. X x Y x Z are not dimensions. Einstein failed to recognize his perspective of Time as Past, Present, and Future combined together into a unit of one is limited because his perspective only occurs when time is a whole which has not been divided or measured. Albert Einstein's perspective on time negates U.S. Census Bureau and U.S. Dept. of Treasury, I.R.S. asking respondent's ages plus dates of birth, researchers surveying age groups of people, and negates all theories of early childhood development because time would be limited to all of time only being a unit of one. We live in divided time.

In that Einstein probably did not have Mr. Binky's Adult Shoppe nearby during his time I will allow that example of vice or the human dimension not be counted against Albert Einstein. However, I am confident examples of the human dimension existed during his time. In thinking about this the Einstein-Szilárd letter sent to United States President Franklin D. Roosevelt on August 2, 1939 supports the Human dimension which means Einstein did not acknowledge the human dimension. Oddly enough Einstein noticed transactions, but he did not expand his perspective to include plant and human transactions. Albert Einstein acknowledged entities but did not expand upon the topic to the extent of including art or non-physics entities as examples of entity in his discussions.

While Albert Einstein's (1920) perspective that time is relative opposes Sir Isaac Newton's (1726/ 1713/ 1687) perspective that time is absolute, both opposing perspectives are correct. Time becomes relative once a data point or point of change has been introduced into time because this introduced point introduces relativity. For example, all time before dinosaurs versus dinosaur-time and later or all time before you were born versus all time after you were born. As well time is absolute until it is measured or has a reference point introduced. To better envision this concept view Figure 13 in Chapter Five and introduce a data point into time and then take out the data point from time.

Place occurs after a data point or measurement is introduced the concept of dimensions cannot support X, Y, and Z as the end-all of all dimensions. If X, Y, and Z were the only dimensions then cases would have been observed with more frequently than the observed four cases or 1.1 percent of total cases with-in the dataset of 300+ published research papers. View Figure 4 in Chapter Two.

My stratified proportionate sampling fails to support the null hypothesis while Einstein's three dimensions support the null hypothesis. The null hypothesis is the academic way of saying there is no relationship between variables which means the data set of peer-reviewed, with references, and English written academic papers support my hypothesis of our six dimensions being time, transaction, entity, place, constants, and human because nothing was observed to negate my hypothesis. Since only four of 347 papers in the data set used Einstein's three-dimensions then the null hypothesis states there is no supported hypothesis for Einstein's three dimensions as all-encompassing dimensions. X x Y x Z is one of many methods to measure place and Place is one of six dimensions.

As universal dimensions, time, transactions, entity, place, and human are specialized enough to be identified as universal dimensions much the same as flour, salt, and pepper are considered kitchen staples. From an overall theoretical perspective all five universal dimensions collapse under the Constant heading because while they are universal dimensions and overall parsimony elicits the simplest format as constant with change. Change can be within or between dimensions. And yes, transactions qualify under this observation because the change within the transaction $4 + 1 = 5$ does exist but does not equal the change within the transaction $1 + 2 = 3$, nor does the change within the transaction F= ma equal the change within the transaction $4 + 1 = 5$.

This research is testable plus the findings are supported. There are six dimensions total: Five dimensions of time, entity, place, constant, and human dimensions, with transactions as an Über-dimension. Mental snap shots or pictures provide static images of the dimensions in action while dynamic images provide for easier observation of role changes.

Multiple Realities and False Dichotomies

This is a good time to discuss multiple realities. These six dimensions support the existence of parallel realities. As mentioned in Chapter One people have physically lived in a perceived three dimensional Einsteinian L x W x D universal reality. You now know you have been physically living in six dimensions. Survey people on their opinions and your findings will support multiple realities. Better research would be to survey people who have not read this book as well survey people who have read this book. Your findings will support multiple realities. Or if you survey people to view art using 20 categories versus surveying people to view art using two or three categories, then your findings will support differing realities for the two sample groups. Our six dimensional reality provides more structure to what you have seen and will see everywhere, every day, and every time.

I set about putting the dimensions into linear sequence and immediately the sociology Structuralism/ Functionalism split from the 60s came to mind. The concern was to explain cultural change and two opposing perspectives stated either the structure of society elicits cultural change or the function of society elicits cultural change. In turn the chicken v egg dichotomy came to mind and merits clarification. The question is 'Which came first the chicken or the egg?' plus in applying this question in this research which dimension would have been the chicken.

Let me answer the chicken and egg question first then we will go back to linear sequence. Recognize the chicken/ egg question is a false dichotomy meaning the answers are not truly limited to two answers rather the parsimony of the question guides one into thinking there are only two viable answers. That sentence is also key to the Structuralist/ Functionalist argument because part of the structuralist perspective, part of the functionalist perspective, and part of other perspectives not included within that dichotomy all contribute towards explaining social change.

The Chicken v Egg question elicits a 2 x 2 table plus since Chicken v Egg asks which one came first then the upper left cell of +Chicken and +Egg, means both the chicken and egg came first, this is not true because both did not occur first. The other three remaining cells +Chicken and –Egg (Chicken came first), –Chicken and +Egg (Egg came first), and –Chicken and –Egg (Neither Chicken nor Egg came first) can be true. The solution could be a pre-chicken came first and that pre-chicken could have been amino acids or an amoeba. See Figure 20.

	+ Egg	- Egg
+ Chicken	Chicken and Egg came first	Chicken came first
- Chicken	Egg came first	Neither Chicken nor Egg came first

Figure 20 Source: Author, False Dichotomy of Chicken v Egg.

Another 2 x 2 table is Thibaut and Kelley's (1959) Social Exchange theory where people update their perspective comparing what they have in relationships against their desires, what they feel they deserve and what is available. Thibaut and Kelley's 2 x 2 table labels fragile relationships which are less likely to last and more likely to break up as the lower right cell with (–) What the person thinks they deserve and (–) What they are receiving (Person is dissatisfied with their relationship and think they deserve better). In contrast strong relationships in the upper left cell are (+) What the person thinks they deserve and (+) What they are receiving (The person is satisfied with their relationship and satisfied with what they are receiving). This 2 x 2 table is clear when the recipient in relationships receive anger and rage and chooses between four options:

- They believe they deserve the abuse and are accepting of their abuse level (+ , +).

- They believe they deserve the abuse and are not accepting of their abuse level (+ , -).

- They do not believe they deserve the abuse and are accepting of their abuse level (- , +).

- The last cell in the 2 x 2 table is they do not believe they deserve the abuse and are looking at the comparative level of alternatives (- , -). The recipient needs safety and the aggressor in this relationship needs professional assistance to un-learn this behavior.

Acknowledgement of false dichotomies allows us to revise and update knowledge and beliefs we hold to be true. Having pro-cessed that we live in a six dimensional universe plus retained your objectivity let us proceed to sequencing these six dimensions. We are not here to argue beliefs, rather we are here to discuss data-driven findings and these six dimensions are supported.

Linear Sequence

Three basic sequence formats are listed from parsimony to complexity. At this point I am not saying any of these sequence formats are supported, rather I am brainstorming so let us assess an order for the occurrence of these dimensions.

Of the three basic concepts for formatting the six dimensions in linear sequence Concept A uses cause and effect plus logic. Austin Hill's (1965) criteria of cause and effect dictates the linear sequence as: Time → Transaction → Entity, Place, and Constants → Human. For Concept A time is independent of other dimensions and can exist by itself surrounded by nothing which will ultimately become the universe. Once that first transaction occurs, then time changes from being 'T minus one' to become 'T zero,' then 'T plus one,' etc.

There is a first transaction and most likely a process of many transactions occurring for an entity to change from zero to plus or into existence. See Figure 21. Once entity exists then place plus constants concurrently come into existence, leaving human dimension at the end. Growth is linear until constants exist. Once constants exist the rotating hypercube comes into operation allowing growth to expand beyond linear to include exponential plus other growth patterns.

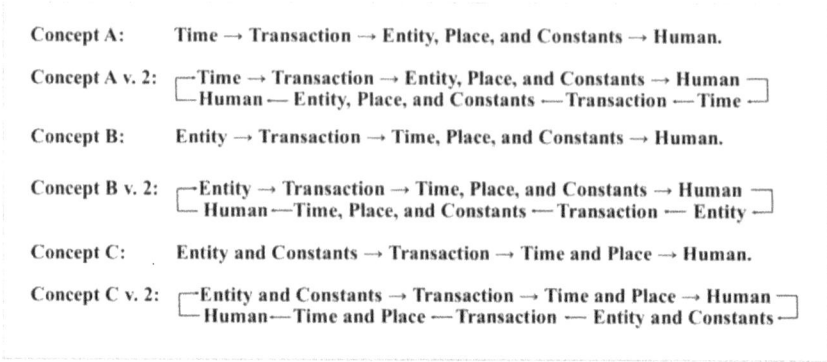

Figure 21 Source: Author, Three Generalized Theories with Secondary Versions of Cosmology Using the Dimensions.

When I scribbled out Concept A I noticed Concept A could be a conceptual flowchart of the Big Bang (Lemaître 1927). Lemaître's focus was measuring the expanding distance between planets and his focus was to work backwards from his known data points to figure out and explain the observed difference and directional transactions evidenced by this expansion. Because of this he never researched nor listed the parts contributing towards making the Big Bang. As well Lemaître never discussed a flowchart for the Big Bang process, nor does Lemaître explain the Big Bang transaction which means Big Bang is more of a working label, similar to 'Black Box Inner-Workings As Yet Unidentified.'

Lemaître labeled the pre-universe expansion transaction Big Bang and this lack of Big Bang definition allows researchers to theorize and test blank spaces in Big Bang transactional process. Perhaps Big Bang was a series of bangs/ transactions or bangs/

transactions of graduated order. As well not all bangs/ transactions have to be kinetic, some could have been electromagnetic. Regardless of those questions there has to have been at least one moment or one change from which before that Big Bang did not exist. Adding to the discussion Seigel (2019) states:

> For the first few tens of millions of years after the Big Bang, gravitation hasn't yet had enough time to work to pull the first neutral atoms together into clumps, meaning we hadn't yet ignited nuclear fusion in them. The only fusion took place during the earliest, hottest, densest stage of the Big Bang, and gave us hydrogen, helium, and not much else.
>
> In fact, after that nuclear fusion took place during the first few minutes of our cosmic history, the Universe needed hundreds of thousands of years to cool by sufficient amounts so that we could stably form neutral atoms. Before that, the photons within it were energetic enough that they continuously knocked every single electron off of whichever atomic nucleus it happened to encounter and bind to.
>
> When the Universe was just a few minutes old, the elements within it were (by weight) about 75% hydrogen, 25% helium, and a tiny fraction of deuterium, helium-3, and lithium. As it cooled over the millennia to come, all of the photons — including the most energetic ones that were primarily responsible for ionization — lost energy. (Seigel, E. 2019)

Seigel (ibid.) supports transactions occurring before entities. These insights allow us to try various sequence and discard parts which do not work.

Concept B's and Concept C's linear sequence differ slightly compared to Concept A, additionally v.2 of all concepts are cyclic. For Concept B Entity is first and creates everything subsequent to itself and the sequence is as follows: Entity → Transaction → Time, Place, and Constants → Human. The benefit to this concept is both Concept A and Concept B support once entity exists, then place, constants, and humans can exist.

Concept C continues the same as Concept B with Entity at the beginning plus Concept C includes Constants at the beginning. After Entity and Constants, transactions occur next, followed by time, place, and humans and is as follows: Entity and Constants → Transaction → Time and Place → Human. The benefit of having Entity plus Constants at the beginning is the rotating hypercube immediately comes into operation allowing growth to expand beyond linear to include exponential plus other growth patterns.

The premise for all v. 2's of Concepts A, B, and C is once the mechanism occurs to create one universe, then the mechanism for serial universes exists. This is because once change has been introduced, then a cycle of change can continue forever eliciting all v. 2's of Concepts A, B, and C. Regardless of the Concept or their v. 2, there are five hot spots with putting these dimensions in order:

- Time,

- Energy,

- Communication with immortal souls,

- Language,

- And the six dimensions.

Hot Spot #1: Negative Time

The downside to Concept B and Concept C is there is no need to have a transaction to create time. The overall model of time allows pre-time (T minus one) to exist and this would relocate time to the left end of Concept B and Concept C's sequence to be concurrent or before entity (entity and constants for Concept C). Plus, if entity exists then place also exists because place comes into existence when entity occurs, which means there is no need for a transaction to create place. A neutral point leaning towards the downside is for Concept B growth is linear until constants exist. Once constants exist the rotating hypercube can come into operation allowing growth to expand beyond linear to include exponential plus other growth patterns.

I understand ancients did not have a NASA program where an engineer states 'T minus three, T minus two, and T minus one' meaning pre-time exists. However, ancients did eat meals so pre-time of 'I am becoming hungry and I need to eat soon' existed and exists. Earlier I mentioned ancients not having a NASA program where an engineer states 'T minus three, T minus two, and T minus one' meaning pre-time or negative time exists. However, ancients did eat meals or sleep so pre-time of 'I am starting to become hungry' or 'I am starting to get drowsy' existed. Ancients just didn't turn the concept around of becoming hungry or tired and then eating or sleeping to recognize the countdown between becoming hungry and eating or becoming tired and sleeping is negative or pre-time and time exists before entity.

Georges Lemaître (1927) wrote the Big Bang was "a day without yesterday" (Midbon 2000) which involved a primal atom, but Lemaître did not write Big Bang was 'T minus one,' 'T is zero,' and 'T plus one,' so he did not recognize negative time. The existence of negative time makes scoffing at Concept A or Concept A v. 2 a difficult argument to support.

Hot Spot #2: Energy Plus Consciousness

Energies can be measured using transactions, but more important is energy is an entity. Applied examples of energy include light switches, lightning bolts, hydro-electric dams, gasoline powered lawn mowers as well as energy of humans. There are various types of energy with quantum mechanics explaining telekinesis transactional processes. Sean McNamara (2016) uses telekinesis to cause foil to rotate when ensconced in an inverted glass vase. The transactional process occurs because the human mind can impact sub-atomic particles when chance processes are involved (Schmidt 1987).

Another type of energy is electromagnetic and this energy facilitates many relationships between and within stars and planets, within and between plants, and within plus between humans (Ronin

Institute n.d. 2018; Grossman 2018; Scott 2018; EM Radiation Research Trust 2017; Gage 2017; Gage 2015; Scott 2015; Tennant 2015; Mallery-Blythe 2014; Markov 2014; Niggli 2014; Tennant 2010; Scott 2007; Thornhill and Talbott 2007; Popp, Chang, Herzog, Yan, and Yan 2002; Cohen and Popp 1997; Popp, Gu, and Li 1994; Popp, Li, and Gu 1992; Ruth 1977; Popp 1975; EMF Research n.d.). Persinger (2011a) tests the parallels between galactic plus extra-galactic electromagnetic fields against the electromagnetic fields of human consciousness concluding, "If consciousness is simply a property of specific electromagnetic patterns and intensities, a variety of these conditions may exist within extracerebral systems." Additionally, in a *Ted* presentation Gage (2017) connects Person A's arm to Person B's arm. Person A opened and closed their fist a couple of times and by using their brain/ electrical impulses Person A is able to make Person B's hand open and close. This research supports consciousness is electromagnetic energy.

Some models of consciousness support the existence of consciousness outside the human body (Gober 2018; Lanza and Berman 2010; Campbell 2005; James 1898) and this type of consciousness is different from Durkheim's (1997/ 1893) collective consciousness of people living within societies having shared beliefs.

Two tangible examples supporting consciousness existing outside our bodies are remote viewers and psychics. Applied parapsychology researchers use human psi or brain energy for their tasks

155

which supports human consciousness existing outside human bodies (Parapsychology Association 2018; IRVA 2018; Swann 2018; Garvin 2017; Radin, Machado, and Zangari 2001; CIA 1998; Puthoff 1996; Radin 1994; CIA 1993; CIA 1988; CIA 1983; CIA 1979; Targ, May, Puthoff, Galin, and Ornstein 1978; Targ and Puthoff 1977; Puthoff and Targ 1974; Osis 1972). My query of 'remote sensing' at CIA Library, FOIA Electronic Reading Room (2019) elicited 12,678 items with an insightful overview of remote viewing (CIA 1988).

While remote viewers primarily perform data collection and analysis for organizations, psychics work in the non-physical realm facilitating communications with physically deceased people. Psychic conversations support an energy-based pre- and post-human life existence which does not require the presence of human bodies. To clarify because physically deceased people who psychics are communicating with are physically deceased and physically-deceased people can still communicate, then physically-dead people are immortal souls or energies existing outside or beyond their deceased body.

Hot Spot #3: Communication

James Van Praagh, Allison DuBoise, and Theresa Caputo have had television series on major television stations involving their ability to regularly communicate with immortal souls. Communicating with immortal souls have become mainstream enough that

Good Housekeeping (Keegan 2018) has an article discussing Caputo performing a psychic reading between a deceased husband/ father with his widow and daughter.

Caputo (Long Island Medium 2017) underwent a brain scan while concurrently performing a psychic reading to a doctor's assistant. The test shows differing use of brain areas when she is channeling compared to when Caputo is not channeling. When remote viewing Ingo Swann's increase in accuracy correlated with 7-Hz EEG spikes and slower wave activity in occipital lobes (Persinger, Roll, Tiller, Koren, and Cook 2002). Gerard Senehi, a mentalist, was tested against a control subject for paranormal abilities. During these tests increased activity in the right parahippocampal gyrus occurred with the mentalist, but not with the control subject (Venkatasubramanian, Jayakumar, Nagendra, Nagaraja, Deeptha, and Gangadhar 2008). These authors conclude the existence of "an association between the right hippocampal system and paranormal phenomena" (ibid.).

While two neurosurgeons (Alexander 2012, Sigmund 2010) wrote about their own near-death experiences with their consciousness continuing into the afterlife an MD writes about her deceased son's continuing experiences in his afterlife. Elisa Medhus MD provides compelling data of psychic-facilitated conversations with her deceased son plus other immortal souls. She co-authored a book with her deceased son using direct communications from her deceased son plus his communications to other family members and psychics (Medhus and Medhus 2013). Additionally, Medhus (2015; ChannelingErik n.d.) writes about her experiences with electrical kitchen appliances both unplugged and plugged-in and telephones starting up and running shortly after her son's death.

Smith-Moncrieffe (2013) uses scientific methods to research psychic mediums and their ability to communicate with immortal souls making the act of dismissing or scoffing at Concept B and C plus their cyclic versions edges upon ignorance. It is clear transactions are occurring from immortal souls to humans or appliances. Regardless of the electromagnetic mechanism for psychic communication the takeaway is communication occurs from immortal souls which means we are immortal.

Hot Spot #4: Language

For language I need to mention some examples. When we were in Australia a possum wandered into our backyard. Except Australian possums are not the same as U.S. possums. Their possums are long bodied and sit on their rear legs in order to hold and eat food with their front paws. When it comes to Colorado, we have coyotes and there have been some coyote poopsie '*scat*' on our dog-walking routes. Using the word scat versus poopsie identifies the group one identifies with the most because I guarantee the composition analysis of coyote fecal matter does not change regardless of how one labels the feces. My third example involves Denver during wintertime.

Denver has Canadian geese and there were some in a nearby field. I pointed to the geese and asked 'He(a)rd of geese?' My husband replied, 'No, gaggle of geese.'

Those examples make me ponder if the energy of Big Bang, which produced and continues to produce through a process of transactions the universe, sun, and us to the extent that we are related albeit distantly to the first time, energies, wavelengths, electrons, etc, is that energy the same energy of Universal Source in that the energy which created Big Bang and subsequent planets is the energy of Universal Source. This means the wording discrepancy between Big Bang and Universal Source is similar to the difference in saying the United States 'tomató' v British 'tomáto,' scat v poopsie, or herd v gaggle.

Envision living 2,000 years ago and you were thinking about cosmology and in this role you are envisioning having a shoebox of nothing. OK, ancients didn't have shoeboxes back then, but it is real tough for me to envision a whole lot of nothing. A shoebox of nothing is easier to envision. So, envision energy which would be written as 'Entity' coming into existence plus that energy created everything subsequent to itself. Because you are envisioning living 2,000 years ago you do not know about our six dimensions since peer-reviewed academic journals with references written in English language does not exist.

It is possible the word for Big Bang is equal to religion and spirituality's Universal Source. It is probable that writers from 2,000 years ago theorized about the Big Bang concept and perhaps the best word for them to choose from their vocabulary of 2,000 years ago was their word for powerful entity. I use the term Universal Source because a concurrent wording throughout all religions does not exist to label a most powerful entity. By writing Universal Source I am labeling the powerful entity discussed within religious and spiritual literature.

I am not saying re-write all science and cosmology books to include all religions because many religious writings proceed beyond cosmology plus include truths, omissions, revisions, bias, psychology and sociology (Greene 2009). In contrast I am asking if your language was limited to only words from 2,000 years ago could the word for something so powerful that it initiated the creation of the universe be Universal Source or Big Bang. We scoff at Aristotle's cosmology of Earth, fire, water, and air plus Einstein's three dimensions of length, width, and depth as being able to describe and quantify reality, yet their ideas were cutting-edge science and technology from their times.

Hot Spot #5: The Six Dimensions

Returning to the six dimensions let us use these dimensions to measure and view finite data points. For example, a single case data point can be you right now using your first-person perspective (human), your location right now (place), you reading this book (transaction and entity), this exact time (time), and whatever constants you want to include.

In contrast when these dimensions are viewed in their entirety without data points or categories dividing them, then these dimensions become all-inclusive to the extent that all of time which has ever and will ever exist plus all of place, and continue through the other dimensions to include all transactions past and future, all entities, all constants, and all perspectives which have ever and will ever exist are contained within Figure 13 in Chapter Five. This grand sum total of all the dimensions becomes the all of ALL and this robust example brings up three important points.

The sequence of events could be that Big Bang is separate from Universal Source plus consciousness and maybe Big Bang does not equal Universal Source. Maybe Big Bang came after Universal Source. For religion or spirituality to participate with cosmology then Universal Source is either Energy or Universal Source is the all of ALL dimensions. Additionally, the singly identified CIA (1988) re-mote sensing overview listed on page 156 has an image of an Einsteinian three-dimensional cube containing data points within the cube plus the paper states:

> Somewhere, perhaps in the unconscious mind, there exists what we will label "The Matrix". The Matrix knows no boundaries and has no limitations – it contains all information about all things. It could be thought of as omnipotent or you could think of it as a data base etc. (CIA 1988, p. 2)

If Universal Source is energy, then CIA's omnipotent Matrix

supersedes Universal Source in omnipotence. This is because the CIA Matrix contains our other five dimensions in addition to energy which is an entity. This premise is supported by pre-time or negative time. Since Figure 13 in Chapter Five plus Chapter Five support infinite constants, then our six dimensions equal the CIA Matrix.

In contrast if Universal Source is the all of ALL dimensions and is not limited, then the CIA Matrix, our Six Dimensions, and Universal Source are the same thing. Along this same line if Universal Source is the all of ALL dimensions and is not limited, then perspectives stating Universal Source creates time are invalid because pre-time or negative time negates this perspective plus Universal Source already is time.

Some people might say the CIA Matrix/ our Six Dimensions/ Universal Source are Akashic records. This insight brings new meaning to U.S. Air Force Technical Applications Center (Romano 2019) motto in their Patrick AFB foyer, "In God We Trust, All Others We Monitor." Because 'CIA Matrix,' 'Six Dimensions,' 'Akashic Records,' and 'Universal Source' could be different names for the same object then substituting these names for God changes the first part of the motto into 'In CIA Matrix We Trust,' 'In Six Dimensions We Trust,' or 'In Akashic Records We Trust' the motto finishing with 'All Others We Monitor.' This foundational insight means all of everybody's thoughts, beliefs, and transactions are already in the Matrix/ Six Dimensions/ Akashic Records/ Universal Source.

In contrast maybe Big Bang came first and then Universal Source. This would mean Big Bang enabled the creation of Universal Source. With this example Universal Source would not have a record of Big Bang. The last option is Big Bang could have occurred simultaneous with Universal Source. Two of these sequences, simultaneous creation and Universal Source existing first, would allow the know-edge of Big Bang to be recorded in Universal Source. If Big Bang existed first, then knowledge of Big Bang would not be recorded in Universal Source. The findings to this research question can be answered by analyzing findings from multiple applied parapsychology researchers (remote viewers and psychics) querying the CIA Matrix/ Six Dimensions/ Universal Source/ Akashic records to identify a record of Big Bang occurring first.

All three sequence candidates plus their cyclic versions support humans existing subsequent to at least one energy transaction. Currently there is not enough data to support Big Bang equals nor does not equal Universal Source. If consciousness created Universal Source, then Durkheim's (1893) collective consciousness comes into play at a grand level.

My last example of a 2 x 2 table is Michelangelo's (1512) Sistine Chapel painting *The Creation of Adam* where God and Humankind are reaching out to establish contact. Is God reaching out to humankind or is humankind reaching out to humankind. The correct answer is it doesn't matter. Treat both answers as equals and apply that knowledge into the real world. It could be our push to understand the Big Bang is misplaced and Big Bang is a scientific event separate from the existence of Universal Source. In conjunction the existence of cerebral plus extra-cerebral consciousness could be of greater importance to our development compared to knowing the steps leading towards the Big Bang.

The sequencing of our six dimensions is becoming interesting and the parsimony of all First versions of the Concepts might be e-nough to justify dismissing all First versions. None of the First versions allowed for serial universe creation. Brown's scientists (Fish 2019) state "Xenon'-124's half-life approximately 18 sextillion years, or 18,000,000,000,000,000,000,000 years- even eclipsing the age of the universe's 13.8-billion-year-old age." Serial universe creation would support this finding.

I keep thinking the mechanics of making the first universe might involve dead ends and false starts, but once the universe has been made then the mechanism exists for repeats. Concept A - First version is too tight of a process and does not explain psychic conversations. I am also tempted to dismiss Concept B - First version and Concept C- First version because they do not allow Time, specifically pre-Time to exist. Both versions of Concept B could be worth dismissing because not allowing constants to occur concurrent to entity limits growth to linear and not exponential growth. What remains is Concept A v. 2 and Concept C v. 2. Concept A v. 2 is a given because of the existence of pre-Time, so potentially the sequence could be a hybrid sequence using Concept A v. 2 and Concept C v. 2. This hybrid sequence elicits pre-time from Concept A v. 2 occurring only once similar to those white pull-tabs one pulls out to establish initial battery and wire contact within widgets. Then implement Concept C v. 2 forever.

Theory of Everything and Unified Fields

While the six dimensions are basic concepts and have been around since early on the grouping, testing, and identifying them as the all-inclusive field-unifying six dimensions which unify under the simple theory of constant with change explains absolutely everything and is new and original and formatting these dimensions with change produces infinite combinations of findings. To paraphrase Noam Chomsky (1957; 1956) from these finite elements, infinite combinations become available.

Minimalists will say one dimension exists and it is change. But while change is the crux and change within and between are powerful measurements alas change must be coupled with something in order for change to exist hence the simplest format is constant. If only constant existed and change did not exist, then we have time without any measurement and you would not be reading this paper. That same argument distinguishes constant as ranking first with change ranking second. Constant with change is the simplest theory and when combined with the crux of this research (quantifying patterned change) and blended with paraphrased Waldo Tobler's First Law of Geography (1970), everything is related to everything else but some things are more related than others, these combined statements form the theory of everything and the six dimensions are the Unified Field Dimensions.

Do not confuse these six dimensions of everything with Kurt Gödel's Incompleteness theory (1931). This is where Gödel's (1931) emphasis on numbers come into play. In paraphrasing Gödel, he said numbers are the answer to everything. He was close because numbers are important in research methods, but numbers do not come first. Gödel did not see that numbers in a mathematical formula have an equal sign in the formula and that equals sign makes his numbers formula a transaction. Gödel believed mathematics supported the finding of a theory of everything and to support this theory everything that existed had to be measured with all measurements combined into one giant formula. By identifying our six dimensions Gödel's requirement of measuring everything becomes un-necessary as evidenced by using statistics to calculate 347 journal articles to generalize findings to the general population. Through identifying our six dimensions the error of Gödel's academic argument becomes clear. These dimensions came first and they are not numbers.

Numbers did not come into play until I was coding and measuring my data. I anticipate Gödel's area for revision was his focus on numbers led into his requirement the answer be a consistent non-trivial mathematical theory. Explained on a conceptual level I am is saying there are six dimensions and by using measurements and not necessarily all dimensions at all times, an infinite number of combinations can occur. This statement is supported by the 93,657 journals with however many number of articles within each journal issue. In contrast to Gödel's theory, my theory does not require identification plus counting of all combinations to know all combinations exist. My data is not inconsistent nor incomplete, rather my data was sampled, tested, and supported.

With Hawking (2002) not being able to format a theory of everything by default Hawking had to support Gödel's Incompleteness Theorem. In contrast these research-supported six dimensions dismiss Gödel's Theorem plus Hawking's cold-shouldering of the existence of a theory of everything through providing unified field dimensions which structure the theory of everything.

Jonas Mureika and Dejan Stojkovic (2011) discuss a vanishing dimension based upon two dimensions of space and following the Big Bang/ transaction more dimensions existed. Their theory can be partially supported in that entity, place, constant, and human became measurable after transaction but their mis-labeled two dimensions (L x W) are a measurement type of the place dimension. Length and width are not independent dimensions. As well with all dimensions being under the heading of constant dimensions, then there is no need for a vanishing dimension.

To clarify many physicists discuss time-space but are incomplete and do not recognize the entity of their discussion, or the entity's transaction within time-space, or theories, or their participation in the conceptualization, viewing, and debriefing to others. Example of these physics theories include General Relativity (Vankov n.d.) with gravity (transaction), space (place), and time, Quantum Field Theory (Dirac 1927) with sub-atomic particles (entities), the interaction between particles and their fields (transactions), plus Grand Unified Theory (Escultura 2008) blends electro-magnetic (entity), weak, and strong interactions (transactions) into a unified coupling constant (transactions).

The above listed theories do not explicitly state and identify our six dimensions all at the same time nor do those theories list our six dimensions as all-inclusive of everything. In contrast the findings from stratified proportionate sampling support our six dimensions are correctly labeled as dimensions and are the most fundamental all-inclusive six dimensions of absolutely everything, everywhere, every time. These six dimensions plus change are our new reality.

This research began with me walking into the campus bookstore and perusing Ritzer's book. Now we know there are six dimensions, we know they are the unified field dimensions, plus these dimensions format the theory of everything. Let's use this new knowledge to do great things.

Let's tap back into the concept of people being icons and put these insights and dimensions into your life skills toolbox to use these tools to repair ourselves, our fragile societies, make our social connections robust, and proceed into the future. Individually and collectively we are traveling our "hero's journey" (Campbell 1949) and these tools will be needed to fulfill our tomorrows. Do not be stymied. There are enough nay-sayers and vendors of doom who make it is easy for us to contribute data towards their predicted failure of mankind by way of us doing nothing. Or we can prove them wrong by surmounting difficulties and accomplishing our true purposes.

References

Adorno, T. and M. Horkheimer. (1972). *Dialectic of Enlightenment.* New York: Seabury Press.

Alexander, E. (2012). *Proof of Heaven: A Neurosurgeon's Journey into the Afterlife.* New York: Simon & Schuster.

Andrews, L. (2014). "Offshore Strip Club in Alaska Waters sees Liquor License Revoked." *Alaska Dispatch News.* December 1, 2014. Accessed from https://www.adn.com on January 5, 2015.

Arensberg, C. (1955). "American Communities." *American Anthropologist,* New Series. 57(6): 1143-1162.

Basker, E. (2005a). "Job Creation or Destruction? Labor Market Effects of Wal-Mart Expansion." *Review of Economics and Statistics.* February 87(1): 174-183.

----------. (2005b). "Selling a Cheaper Mousetrap: Wal-Mart's Effect on Retail Prices" *Journal of Urban Economics.* 58(2): 4-27.

BBC. (2010). "The History of Hamleys – London's Famous Toy Shop." *BBC.* February 11, 2010. Accessed from http://news.bbc.co.uk/local/london/hi/people_and_places/history/newsid_8510000/8510277.stm on June 7, 2019.

Berger, P. and T. Luckman. (1966). *The Social Construction of Reality.* New York: Penguin Books.

Bielefeldt, J., Bozek, M., Grosshuesch, D., Murphy, R., Rosenfield, L., and R. Rosenfield. (2003). "Comparative Relationships Among Eye Color, Age, and Sex in Three North American Populations of Cooper's Hawks." *Wilson Bulletin.* Sept: 225.

Blau, P. (1964). *Exchange and Power in Social Life.* New York: Wiley.

Campbell, J. (1949). *Hero With A Thousand Faces.* New York: Pantheon Books.

Campbell, T. (2005). *My Big T.O.E.: A Trilogy Unifying Philosophy, Physics, and Metaphysics: Awakening, Discovery, Inner Workings.* Lightning Strike Books.

Chomsky, N. (1957). *Syntactic Structures*, Mouton, The Hague.

----------. (1956). "Three Models for the Description of Language." *IRE Transactions on Information Theory*, IT-2:3, 113-124.

CIA. (2019). Central Intelligence Agency, Library, Electronic Reading Room, FOIA. Accessed from https://www.cia. gov/library/readingroom/search/site/remote%20viewing on January 4, 2019.

----------. (1998). *"Proposal for a One-Year Research Study to Determine The Role, If Any, Of Information Obtained From Psychics In Fulfilling CIA's Missions."* Central Intelligence Agency, Library, Electronic Reading Room. Accessed from https://www.cia.gov/library/readingroom/docs/CIA-RDP96-00787R000200020 021-7.pdf on January 14, 2019.

----------. (1993). *"Standard Remote Viewing (RV) Procedures: Local Targets."* Puthoff, H. and Targ, R. Central Intelligence Agency, Library, Electronic Reading Room. Accessed from https://www.cia.gov/library/readingroom/docs/CIA-RDP96-00788 R001300050001-3.pdf on January 22, 2019.

----------. (1988). *"Coordinate Remote Viewing (Theory and Dynamics)."* Central Intelligence Agency, Library, Electronic Reading Room. Accessed from https://www.cia.gov/library/readingroom/docs/CIA-RDP96-00789R001300010001-6.pdf on January 4 2019.

----------. (1983). *"Coordinate Remote Viewing Technology (CRV) 1981-1983 Three Year Project Draft Report."* Central Intelligence Agency. Accessed from https://www.cia.gov/library/readingroom/docs/CIA-RDP96-00788R00180010000 1-2.pdf on January 14, 2019.

----------. (1979). *"Standard Remote Viewing (RV) Procedures: Local Sites."* Puthoff, H., Targ, R., and E. May. Central Intelligence Agency, Library, Electronic Reading Room. Accessed from https://www.cia.gov/library/readingroom/docs/CIA-RDP 96-00788R002000240 029-4.pdf on January 14, 2019.

Clarage, M. (2018). "Special Feature: New Views of Interstellar Medium – Michael Clarage." *ThunderboltsProject*. Accessed from https://www.youtube.com/wat ch?v=CPQRv_6J8WM on December 5, 2018.

Cohen, S., and F. Popp. (1997). "Biophoton emission of the human body." *Journal of Photochemistry and Photobiology B Biology*. 40(2): 187-189.

Constance, D., Harvel, D., and A. Prelog. (2017). "Tennessee Whiskey as Terroir?" Paper presented at annual meeting of the Southwestern Sociological Association. Austin, TX. April 14, 2017.

DeMers, M. (2002). *GIS Modeling in Raster*. New York: John Wiley & Sons.

Dirac, P. (1927). "The Quantum Theory of Emission and Absorption of Radiation." *Proceedings of the Royal Society of London*. A (114): 243-256.

Durkheim, E. (1995/ 1912). *The Elementary Forms of Religious Life* (trans Karen Fields). Translation of Les Formes élementaires de la vie religieuse: Le système totémique en Australie, 1912. New York: Free Press.

----------. (1997/ 1893). *The Division of Labour in Society* (trans W. Halls, intro. L. Coser). New York: Free Press, 1997

Einstein, A. (1961/ 1952). *Relativity: The Special and the General Theory*. New York: Crown Trade Paperbacks.

----------. (1920). "Time, Space, and Gravitation." *Science*. 51(1305): 8-10.

----------. (1916). "Review Article on General Relativity." *Annalen der Physik*. 49: 769-822.

----------. (1905). "On the Electrodynamics of Moving Bodies." *Annalen der Physik*. 17: 891.

EM Radiation Research Trust. (2017). *"Dr. Erica Mallery-Blythe – Electromagnetic Radiation, Health and Children 2014."* Accessed from https://www.radiation research.org/articles/dr-erica-mallery-blythe-electromagnetic-radiation-health-and-children-2014/ on December 21, 2018.

EMF Research. (n.d.). *"Dr Erica Mallery-Blythe – Electromagnetic Radiation, Health + Children."* Accessed from www.emfre

search.com/dr-erica-mallery-blythe-electromagnetic-radia
tion-health-children/ on December 21, 2018.

Escultura, E. (2008). "The Grand Unified Theory." *Nonlinear An-alysis*. 69: 823-831.

Fish, T. (2019). "Dark Matter: 'Rarest Event EVER' Recorded by Scientists-Lasting 18 Sextillion Years." *Express*. Published on May 27, 2019. Accessed from https://www.express.co. uk/news/science/1124073/dark-matter-detector-xenon-124-rarest-event-ever-recorded on June 7, 2019.

Foucault, M. (2001). *Power: Essential Works of Foucault, 1954-1984, Volume 3.* (ed J. Faubion, R. Hurley). The New Press.

Gage, G. (2017). "Electrical Experiments With Plants that Count and Communicate." *Ted*. November 2017. Accessed from https:// www.youtube.com/watch?v=pvBlSFVmoaw on May 21, 2019.

----------. (2015). "How to Control Someone Else's Arm with Your Brain." *Ted*. April 2015. Accessed from https://www.you tube.com/watch?v=rSQNi5sAwuc on May 21, 2019.

Garvin, G. (2017). "CIA files reveal how U.S. used psychics to spy on Iran." Miami Herald. Accessed from https://www.miami herald.com/news/nationworld/national/article131827589.htm l on December 4, 2018.

Geertz, C. (1973). *The Interpretation of Cultures*. New York: Basic Books Inc.

Glass, R. (1955). "Urban Sociology in Great Britain: a Trend Report." *Current Sociology*. 4: 1-12.

Glazer, N. and D. Moynihan. (1970). *Beyond the Melting Pot.* Cambridge, MA: The MIT Press.

Gleick, J. (1987). *Chaos: The Making of a New Science.* Viking Adult.

Gobe, M. and S. Zyman. (2001). *Emotional Branding: The New Paradigm for Connecting Brands to People.* Watson-Guptill Publications.

Gober, M. (2018). *An End to Upside Down Thinking.* Waterside Productions.

Gödel, K. (1931). "Über Formal Unentscheidbare Sätze der Principia Mathematica und Verwandter Systeme, I." *Monatshefte für Mathematik und Physik.* 38(1): 173-198.

Goffman, E. (1959). *The Presentation of Self in Everyday Life.* NY: Random House Publishing.

----------. (1963). *Stigma: Notes on the Management of Spoiled Identity.* NY: Random House Publishing.

Gramsci, A. (1975/ 1932). *Letters from Prison: Antonio Gramsci.* (ed Lynne Lawner). New York: Harper Colophon.

Greene, R. (2009). "Oldest Bible Goes Online." *CNN US Edition.* July 6, 2009. Accessed from http://www.cnn.com/2009/WORLD/europe/07/06/ancient.bible.online/index.html on January 23, 2019.

Grossman, D. (2018). "Government Accidentally Releases Documents on "Psycho-Electric" Weapons." *Popular Mechanics.* April 19, 2018. Accessed from *https://www.popularmechanics.com/military/weapons/a19855256/muckrock-foia-psycho-electric-weapons/* on April 21, 2018.

Halbwachs, M. (1980/ 1950). *The Collective Memory* (trans F. Ditter Jr., V. Yazdi Ditter, 1980). New York: Harper Colophon Books.

Hamilton, R. and A. Chernev. (2010). The Impact of Product Line Extensions and Consumer Goals on the Formation of Price Image. *Journal of Marketing Research.* 47:1. Accessed from http://www.chernev.com/research/articles/The_Impact_of_ Product_Line_Extensions_and_Consumer_Goals_on_the_ Formation_of_Price_Image_2010.pdf on October 12, 2013.

Hawking, S. (2002). "Godel and the End of Physics." Accessed from http://www.damtp.cam.ac.uk/events/strings02/dirac/hawking / on November 30, 2018.

Higgins, J. (2007)."The St. John's Fire of 1892." *Newfoundland and Labrador Heritage.* Accessed from https://www.heritage.nf-.ca-articles/politics/st-johns-fire-1892.php on July5, 2019.

Hicks, M. and K. Wilburn. (2001). "The Regional Impact of Wal-Mart Entrance: A Panel Study of the Retail Trade Sector in West Virginia." *Review of Regional Studies.* 31(3): 305-313.

Hill, A. (1965). "The Environment and Disease: Association or Causation?" *Proceedings of the Royal Society of Medicine.* 58: 295-300.

Holt, D. (2004). *How Brands Become Icons.* Mass: Harvard Business School Press.

IRVA. (2018). *IRVA International Remote Viewing Association.* Accessed from https://www.irva.org/index.html on December 4, 2018.

Jacobs, J. (1961). *The Death and Life of Great American Cities*. New York: Knopf Publishing Group.

James, W. (1898). *Human Immortality: Two Supposed Objections to the Doctrine*. Houghton Mifflin.

Jobim, A. (2008/ 1959). "One Note Samba." *One Note Samba*. Universal Portugal, 2008: 11.

Jones, C. (2017). "Sears 4Q Loss is the Latest in a Years-Long String." *USA Today*. March 9, 2017. Accessed from https://www.usatoday.com/story/money/2017/03/09/sears-kmart-earnings/98945356/ on March 10, 2017.

Katz, D. and R. Kahn. (1966). *The Social Psychology of Organization*. New York: Wiley.

Keegan, K. (2018). "'Long Island Medium' Theresa Caputo Makes Mom and Daughter Hysterical With Reading." *Good Housekeeping*. November 16, 2018. Accessed from https://www.goodhousekeeping.com/life/entertainment/a25138290/long-island-medium-season-13-episode-7-sneak-peek/ on November 19, 2018.

Kotler, P. (2005). *Brand Sense: Build Powerful Brands Through Touch, Taste, Smell, Sight, and Sound*. New York: Free Press.

Lanza, R. and B. Berman. (2010). *Biocentrism: How Life and Consciousness are the Keys to Understanding The True Nature of the Universe*. BenBella Books.

Lawrence, S. (2008). *Businesses as Cultural Icons: Their Application towards Understanding Urban Morphology* (Doctoral dissertation). University of New Orleans.

Lawrence, S. and J. Wildgen. (2012). Manifold Destiny: Migrant Deaths and Destinations in the Arizona Desert. *Growth and Change.* 43(3): 482-504.

Lebhar, G. (1953). *Chain Stores in America, 1859-1950.* New York: Chain Store Publishing Company.

Lee, T., Dumas, T., and J. Jevnikar. (1983). "Comparison of the Effects of Glyphosate and Related Compounds on Indole-3-Acetic Acid Metabolism and Ethylene Production in Tobacco Callus." *Pesticide Biochemistry and Physiology.* 20: 354359.

Lemaître, G. (1927). Annales de la Société Scientifique de Bruxelles, Series A47: 49-59.

Lo, C., and A. Yeung. (2002). *Concepts and Techniques of Geographic Information Systems.* New York: Prentice Hall.

Long Island Medium. (2017). "What Makes Theresa Caputo's Brain Unlike Others? Long Island Medium." *YouTube.* Published on April 27, 2017. Accessed from https://www.youtube.com/watch?v=E8qj0EPLwnQ on November 2, 2018.

Lukács, G. (1968/ 1922). *History and Class Consciousness.* Cambridge: MIT Press.

Macys. (2017). "Macy's Inc. Reports Fourth Quarter and FY2016 Results." Press Release. Accessed from https://www.macys inc.com/news-media/press-releases/detail/714/macys-inc-re ports-fourth-quarter-and-fy2016-results on February 23, 2017.

Mallery-Blythe, E. (2014). "Electromagnetic Hypersensitivity: A Summary Submitted to EESC." Accessed from https://phire medical.org on November 30, 2018.

Markov, M. (2014). "Electromagnetic Fields and Life." *Journal of Electrical & Electronic System.* 3(1).

Marx, K. (1992). *Capital: Volume 1: A Critique of Political Economy* (trans B. Fowkes) Penguin Publishing Group.

Mauss, M. (1990/ 1950). *Sociologie et anthropologie,* 1950. *The Gift* (trans W. Halls, 1990). London: Routledge.

McHarg, I. (1995/ 1969). *Design With Nature.* Wiley Publishing.

McMillan, D. and D. Chavis. (1986). "Sense of Community: A Definition and Theory." *American Journal of Community Psychology.* 14(1): 6-23.

McNamara, S. (2016). "Telekinesis with a MASSIVE target by Sean McNamara of MindPossible." *YouTube.* Published on August 6, 2016. Accessed from https://www.youtube.com/watch?v=YRfcj0FEHqE on August 12, 2016.

Medhus, E. (2013). *My Son and the Afterlife: Conversations from the Other Side.* Atria Books/ Beyond Words.

----------. (2018). *Channeling Erik: Conversations With My Son in the Afterlife.* Accessed from https://channelingerik.com/ on October 12, 2018.

Medhus, E., and E. Medhus. (2015). *My Life After Death: A Memoir from Heaven.* New York: Atria Books/ Beyond Words.

Mellor, R. (1975). "Urban Sociology in an Urbanized Society." *The British Journal of Sociology.* 26(3): 276-293.

Michaelangelo. (1512). *The Creation of Adam.* Paint on Plaster. Sistine Chapel in Vatican City.

Midbon, M. (2000). "'A Day Without Yesterday': Georges Lemaitre & the Big Bang." *Catholic Education Resource Center.* Published in 2000. Accessed from https://www.catholic education.org/en/science/faith-and-science/aday-without-yes terday-georges-lemaitre-amp-the-bigbang.html on November 26, 2018.

Mumford, L. (1938). *The Culture of Cities.* New York: Harcourt, Brace and Company.

Mureika, J., and D. Stojkovic. (2011) "Detecting Vanishing Dimensions via Primordial Gravitational Wave Astronomy." *Physical Review Letters.* 106: 101101.

Newton, I. (1726/ 1713/ 1687). *Philosophiae Naturalis Principia Mathematica.* (trans A. Motte, 1729). London: Jussu So cietatis Regiæ ac Typis Joseph Streater.

Niggli, H. (2014). "Biophotons: Ultraweak Light Impulses Regulate Life Processes." *Gerontology & Geriatric Research.* 3:2.

Oldenburg, R. (1999). *The Great Good Place.* Marlowe & Co.

Osis, K. (1972). *"New ASPR Search on Out-of-the-Body Experiences."* ASPR Newsletter. 14: Pp. 2, 4. Accessed from https://www.cia.gov/library/readingroom/docs/CIA-RDP96-00787R000400100012-6.pdf on January 22, 2019.

Parapsychology Association. (2018). *Parapsychology Association.* Accessed from https://www.miamiherald.com/news/nation-world/national/article131827589.html on December 4, 2018.

Persinger, M. (2011a). "Electromagnetic Bases of the Universality of the Characteristics of Consciousness: Quantitative Support." *Journal of Cosmology, 2011, Vol. 14.* Accessed from http://www.nrgarchive.gdk.mx/2011-Persinger-Journal-of-

Cosmology-Electromagnetic-bases-of-the-universality-of-the-characteristics-of-consciousness.pdf on January 10, 2019.

----------. (2011b). "No More Secrets by Michael Persinger." *You tube.* Published on March 30, 2011. TVO Big Ideas presents Michael Persinger, winner of the 2007 Best Lecturer Competition, on Just Suppose You Could Know What Others Are Thinking: No More Secrets. Accessed from https://www.youtube.com/watc h?v=9l6VPpDublg#action=share on June 5, 2019.

Persinger, M., Roll, W., Tiller, S., Koren, S., and C. Cook. (2002). "Remote Viewing with the Artist Ingo Swann: Neuropsychological profile, electroencephalographic cor-relates, magnetic resonance imaging, and possible mech-anism." *Perceptual and Motor Skills.* 94(3): 927-949.

Picasso, P. (1957). Dog. Accessed from http://www.pablo picasso.net /dog/ on April 12, 2017.

----------. (1958). Bouquet of Peace. Accessed from https://www. masterworksfineart.com/artists/pablopicasso/lithograph/bouq uet-of-peace-1958-2/id/w-2371 on April 12, 2017.

Pine, B., and J. Gilmore. (1999). *The Experience Economy: Work Is Theatre & Every Business a Stage.* Mass: Harvard Business School Press.

Popp, F. (n.d.). *International Union of Medical and Applied Bio-electrography.* Accessed from https://www.iumab.org on No-vember 30, 2018.

Popp, F., Li, K. and Q. Gu. (1992). "Recent Advances in Biophoton Research and Its Application." *World Scientific.* 1-18.

Popp, F., Gu, Q., and K. Li. (1994). "Biophoton Emission: Experimental Background and Theoretical Approaches." *Modern Physics Letters B*. 8: 21-22.

Popp, F., Chang J., Herzog, A., Yan, Z., and Y. Yan. (2002). "Evidence of Non-Classical (Squeezed) Light in Biological Systems." *Physics Letters A*. 293(1-2): 98-102.

Powell, W. and P. DiMaggio, eds. (1991). *The New Institutionalism in Organizational Analysis*, Chicago, IL/London, Chicago University Press.

Puthoff, H. (1996). "CIA-Initiated Remote Viewing At Stanford Research Institute." Accessed from https://web.archive.org/web/20100104220426/http://www.biomindsuperpowers.com/Pages/CIA-InitiatedRV.html on January 22, 2019.

Puthoff, H. and R. Targ. (1974). "Physics, Entropy and Psychokinesis." L. Oteri (ed.). Quantum Physics and Parapsychology: Proceedings of an International Conference Held in Geneva, Switzerland, August 26-27, 1974. Pp. 129-150. New York: Parapsychology Foundation.

Radcliff-Brown, A. (1931). "The Social Organization of Australian Tribes." Oceania Monographs, no. 1. Melbourne: Macmillan.

----------. (1952). Structure and Function in Primitive Society. New York: Free Press.

Radin, D. (1994). "Environmental Modulation and Statistical Equilibrium in Mind-Matter Interaction." *Subtle Energies & Energy Medicine.* 4(1): 1-30. Accessed from http://journals.sfu.ca/seemj/index.php/seemj/article/view/158/123 and http://deanradin.com/articles/1993%20equilibrium.pdf on January 22, 2019.

Radin, D., Machado, F., and W. Zangari. (2001). "Effects of Distant Healing Intention Through Time & Space: Two Exploratory Studies." *Subtle Energies & Energy Medicine*. 11(3): 207-239. Accessed from http://journals.sfu.ca/seemj/index.php/seemj/article/view/309/272 on January 22, 2019.

Ronin Institute. (n.d.). "Michael Clarage." Accessed from http://ronininstitute.org/research-scholars/michael-clarage/ on December 3, 2018.

Ritzer, G. (2004). *The Globalization of Nothing*. Thousand Oaks: Sage Publications.

----------. (1993). *The McDonaldization of Society: An Investigation into the Changing Character of Contemporary Social Life.* Pine Forge Press Publication.

Romano, S. (2019). "New Place to Call Home." *U.S. Air Force Civil Engineer, Photos.* Accessed from https://www.afcec.af.mil News/Photos/ig photo/2000858543/ on January 15, 2019.

Ruth, B. (1977). "Experimenteller Nachweis Ultraschwacher Photonenemission aus Biologischen Systemen." Doctoral Thesis, University of Marburg, Germany.

Sarlo, C. (1992). *Poverty in Canada.* Vancouver: The Fraser Institute.

---------. (2001). *Measuring Poverty in Canada.* Critical Issues Bulletin. Vancouver: The Fraser Institute.

---------. (2006a). *Comparing Measures of "Poverty."* Fraser Forum. Vancouver: The Fraser Institute.

---------. (2006b). *Poverty in Canada, 2006 Update.* Fraser Alert. Vancouver: The Fraser Institute.

Schmidt, H. (1987). "The Strange Properties of Psychokinesis." *Journal of Scientific Exploration.* 1(2): 103-118.

Scott, D. (2018). "Birkeland Currents and Dark Matter." *Progress in Physics.* 14(2): 57-62. Accessed from http://www.ptep-on line.com/2018/PP-53-01.PDF on December 13, 2018.

----------. (2015). "Birkeland Currents: A Force-Free Field-Aligned Model." *Progress in Physics.* 11(2): 167-179. Accessed from http://electric-cosmos.org/ on December 13, 2018.

----------. (2007). "Real Properties of Electromagnetic Fields and Plasma in the Cosmos." *IEEE Transactions on Plasma Science.* 35(4): Pp. 822-827. Accessed from http://electric-cosmos.org/IEEE-TransPlasmaSci-Scott-Aug2007.pdf on December 14, 2018.

Seigel, E. (2019). "Scientists Didn't Really Find The First Molecule In The Universe." *Forbes.* April 19, 2019. Accessed from https://www.forbes.com/sites/startswithabang/2019/04/19/sci entists-didnt-really-find-the-first-molecule-in-the-universe/#-45c6c90a5ee1 on April 29,2019.

Sigmund, R. (2010). *My Time in Heaven: A True Story of Dying and Coming Back.* Anchor Distributors.

Smith-Moncrieffe, D. (2013). *Medium7: Evidence of the Afterlife and Predictions.* Bloomington: iUniverse.

Swann, I. (2018). *Everybody's Guide to Natural ESP: Unlocking the Extrasensory Power of Your Mind.* Swann-Ryder Produc-tions.

Targ, R., May, E., Puthoff, H., Galin, D., and R. Ornstein. (1978). *"Sensing of Remote EM Sources (Physiological Correlates)."*

SRI International Final Report on Naval Electronics Systems Command Project N00039-76-C-0077, covering the period November 1975 - to October 1976. Click on "See Other Formats" and on lower right are download options. Choose PDF w Text. 44 pages. Prepared for Naval Electronic Systems Command, Washington D.C. Accessed from https://archive.org/details/CIA-RDP96-00788R001300350001-0. https://ia801205.us.archive.org/19/items/CIA-RDP96-00788 R001300350001-0/CIA-RDP96-00788R0013003500010_ text.pdf on January 22, 2019 .

Targ, R. and H. Puthoff. (1977). *Mind-Reach: Scientists Look at Psychic Ability*. Jonathan Cape Ltd.

Taylor, F. (1911). *The Principles of Scientific Management*. New York: Harper and Brothers.

Tennant, J. (2015). *Healing is Voltage: Cancer's On/ Off Switches: Polarity*. CreateSpace Independent Publishing Platform.

----------. (2010). *Healing is Voltage: The Handbook*. CreateSpace Independent Publishing Platform.

Thibaut, J. and H. Kelley. (1959). *The Social Psychology of Groups*. New York: Wiley.

Thornhill, W., and D. Talbott. (2007). *The Electric Universe*. Mikamar Publishing.

Tobler, W. (1970). "A Computer Movie Simulating Urban Growth in the Detroit Region." *Economic Geography*. 46:234-240.

Tönnies, F. (1957/ 1887). *Community and Society* (trans C. Price Loomis, 1957). East Lansing: Michigan State University Press. Translation of *Gemeinschaft und Gesellschaft*, Leipzig: Fues's Verlag, 1887.

Tuder, S. (2015). "The Quest to Save the First Taco Bell From Destruction." Accessed from http://abcnews.go.com/Life style/questsavetacobelldemolition/story?id=29762492 on April 1, 2015.

Tylor, E. (1924/ 1871). *Primitive Culture*: 1. 2 vols. 7th ed. New York: Brentano's.

Veblen, T. (2011/ 1899). *The Theory of The Leisure Class.* New York: Oxford University Press Inc.

Vankov, A. (n.d.). Einstein's Paper: "Explanation of the Perihelion Motion of Mercury from General Relativity Theory." Accessed from http://www.gsjournal.net/old/eeuro/vankov. pdf on October 30, 2014.

Venkatasubramanian, G., Jayakumar, P., Nagendra, H., Nagaraja, D., Deeptha, R., and B. Gangadhar. (2008). "Investigating Paranormal Phenomena: Functional Brain Imaging of Telepathy." *International Journal of Yoga.* 1(2): 66-71.

Ziegelman, J. (2011). *97 Orchard: An Edible History of Five Immigrant Families in One New York Tenement.* New York: Harper Paperbacks.

About the Author

Prior to being awarded her PhD in Urban Studies from University of New Orleans Stephanie Lawrence received her BS in Sociology from Texas A&M University and MA in Sociology from University of New Orleans. Additionally, she co-authored "Manifold Destiny: Migrant Deaths and Destinations in the Arizona Desert" with John Wildgen (Lawrence, S. and J. Wildgen. [2012]. "Manifold Destiny: Migrant Deaths and Destinations in the Arizona Desert," *Growth and Change*. 43 (3): 482-504, September). She also co-presented research on the best city governance style at UAA (Urban Affairs Association 2004).

Stephanie Lawrence PhD is the subject matter expert on *Our Six Dimensions* and presented her research findings examining the six dimensions plus businesses as urban cultural icons at three conferences: AAG (American Association of Geographers) 2017, SSSA (Southwest Social Science Association) 2017, and UAA (Urban Affairs Association) 2017, receiving good reviews.

The above paragraphs list her serious credentialing. Her life credentialing includes having two children, two Weiner dogs, and this year is her 39th year of marriage. Early on in their marriage they agreed to follow her husband's career. Since he was a petroleum engineer they relocated throughout southern and western United States, plus a stint in Australia. This is important because the businesses she chose for measuring businesses as urban cultural icons came from living in the U.S. South and West. Some of the businesses she visited include Buffalo Bill's Irma Hotel in WY (named for his daughter), Brookville Hotel KS, Bad Ass Coffee TX, Foss Drugs CO, and Bellagio, Tropicana, plus Caesar's Palace NV.

www.ingramcontent.com/pod-product-compliance
Lightning Source LLC
Chambersburg PA
CBHW050650270326
41927CB00012B/2964